Circular Bioeconomy

Fazia Adyani Ahmad Fuad · Faridah Yusof ·
Amanatuzzakiah Abdul Halim · Farah Ahmad ·
Yusilawati Ahmad Nor
Editors

Circular Bioeconomy

Towards a Sustainable Future

Springer

Editors
Fazia Adyani Ahmad Fuad
Department of Chemical Engineering and Sustainability
Kulliyyah of Engineering
International Islamic University Malaysia (IIUM)
Selangor, Malaysia

Faridah Yusof
Department of Chemical Engineering and Sustainability
Kulliyyah of Engineering
International Islamic University Malaysia (IIUM)
Selangor, Malaysia

Amanatuzzakiah Abdul Halim
Department of Chemical Engineering and Sustainability
Kulliyyah of Engineering
International Islamic University Malaysia (IIUM)
Selangor, Malaysia

Farah Ahmad
Department of Chemical Engineering and Sustainability
Kulliyyah of Engineering
International Islamic University Malaysia (IIUM)
Selangor, Malaysia

Yusilawati Ahmad Nor
Department of Chemical Engineering and Sustainability
Kulliyyah of Engineering
International Islamic University Malaysia (IIUM)
Selangor, Malaysia

ISBN 978-981-97-7009-0 ISBN 978-981-97-7010-6 (eBook)
https://doi.org/10.1007/978-981-97-7010-6

This Springer imprint is published by the registered company Springer Nature Singapore Pte Ltd.
The registered company address is: 152 Beach Road, #21-01/04 Gateway East, Singapore 189721, Singapore

Preface

The concept of circular bioeconomy signifies a fundamental change in the way we view and utilise our natural resources. This book explores the intertwined fields of biology, chemistry, economics, and technology, all of which come together to reset how we think about production, consumption, and waste. The urgency to address climate change, resource depletion, and environmental degradation forces us to reconsider conventional linear models of extraction, use, and disposal. Circular bioeconomy, a regenerative system that manages biological resources sustainably and cycles them back and forth in a closed-loop cycle, offers an alternative to the traditional route. As a result, ecosystems are restored, waste is turned into a useful input, and communities live in a sustainable environment.

This book explores the multifaceted dimensions of the circular bioeconomy, which is presented in twelve (12) chapters, focussed on five (5) sections, namely, Biomaterials, Environment, Energy, Pharmaceutical, and Enzyme Technology. It portrays some of the areas of research which have been conducted or being conducted by our research members, linking circular bioeconomy to many of the United Nation's Sustainable Development Goals (SDGs). Each chapter dives into important facets of this paradigm shift, ranging from the basic ideas of biological processes and ecosystem dynamics to the state-of-the-art technology and methods, allowing bio-based advancements.

It is important to emphasise that circular bioeconomy is the viable path towards sustainability, rather than just a theoretical concept. In order to realise its potential, joint actions from various stakeholders; scientists, engineers, entrepreneurs, policy-makers, and citizens are needed from a variety of disciplines and sectors. Through the chapters, we invite all readers to embark on a journey to discover the prospect of circular bioeconomy through its application in various areas and industries. This book will provide the opportunity for all readers to delve into the true potential of circular bioeconomy, towards a sustainable future.

Selangor, Malaysia
July 2024

Fazia Adyani Ahmad Fuad
Chief Editor

Contents

Enzyme Technology

Recovery of Bioactive Peptides from the Agri-Industrial Wasteful Eggshell Membranes

Aina Basyirah Ahmad Razali and Faridah Yusof

Abstract The increasing trends of chicken egg consumption have led to the generation of high amounts of eggshells and their membranes as part of the agri-industrial waste. Due to its unique organic characteristics and valuable composition, the waste eggshell membrane (ESM) has been exponentially utilized within the nutraceuticals and pharmaceutical industry to produce natural bioactive peptides. Within the framework of circular economy, this chapter explores the overview of the chicken ESM, including their functional characteristics and various applications. This is followed by the recovery processes of bioactive peptides from ESM to maximize this valuable resource utilization and minimize the waste management issues in the production of environmental-friendly biomaterials.

1 Introduction

Since ancient times, humans have been consuming animals' eggs as a staple food due to their cheap and readily available source of protein with chicken eggs being the most favored. In 2020, more than 86.67 million metric tons of chicken eggs were produced globally with a weight of approximately 50 g to 70 g per egg (Statista Research Department 2022a). According to the Consumers' Association of Penang (2022), annually, 370 chicken eggs per capita are consumed on average in Malaysia, which equals to the market demand of around 30 million eggs daily. It is recorded that Malaysia had been contributing to almost 1% of global egg production in 2020 (Statista Research Department 2022b).

In reference to the study of the oology of the eggs' anatomy and physiology, a typical egg consists of a protective eggshell (ES) layer, its ESM, the albumen known as egg white, and the vitellus known as the yolk (Mittal et al. 2016). One-tenth of the total weight of an egg is represented by the ES and the ESM and these are generated as agri-industrial waste (Shi et al. 2021). The global production of ES and associated

A. B. A. Razali · F. Yusof (✉)
Department of Chemical Engineering & Sustainability, Kulliyyah of Engineering, International Islamic University of Malaysia (IIUM), Jalan Gombak, 53100 Kuala Lumpur, Malaysia
e-mail: yfaridah@iium.edu.my

F. Adyani Ahmad Fuad et al. (eds.), *Circular Bioeconomy*,
https://doi.org/10.1007/978-981-97-7010-6_1

ESM, considered a major waste from the food industry, reaches approximately 9 million tons per year, constituting 11% of the global egg production of 80 million metric tons (FAO, 2020). The increase in these wastes of ES and ESM, generated from the poultry, domestic, and food-manufacturing industries had contributed to the flooding of landfills, raising the cost of waste disposal and management with several rising environmental concerns.

Due to the organic characteristics of ESM which enhance enzymes and protein formation as opposed to the inorganic calcified ES, the usage of ESM had been implemented as a source of biomaterial in a vast area of engineering fields including materials, electrical, chemical, nutraceutical, cosmetic, environmental, and biomedical fields (Baláž 2014). This promotes the transformation of egg production by-products into wealth to support the framework of bioeconomy and enhance the circularities of the economy by maximizing the use of waste resources and minimizing the issues of waste management. This idea is also enhanced by the United Nations' Sustainable Development Goal 12 of Responsible Consumption and Production which targets sustainable consumption and production patterns by halving global food waste by 2030 (United Nations 2022).

Within the framework of circular economy, this chapter explores the overview of the chicken ESM, including their functional characteristics, and their various applications followed by the recovery processes of bioactive peptides from ESM to potentially expand the nutraceutical and pharmaceutical segments.

2 Eggshell Membrane (ESM)

The calcareous egg is produced by all birds and most reptiles. The current understanding of ES formation and mineralization is mainly based on intensive studies of one species, the domesticated chicken *Gallus gallus*. The chicken ES has evolved to withstand physical and pathogenic challenges from the external environment (Hincke et al. 2012). The ESM, located between the albumen and the inner ES surface is approximately 100 mm thick (Jain & Anal 2016). The proteinaceous ESM, which makes up around 1% of the egg's weight, firmly adheres to the ES's inner surface, whereas the ES, which makes up around 10% of the egg's weight, forms the tough calcareous outer layer (Kulshreshtha et al. 2022). During ES calcification in the egg formation process, the ESM acts as a polymeric fibrous meshwork that facilitates biomineralization while blocking mineralization toward the inner egg white.

2.1 *Physical Properties of ESM*

According to Kadir et al. (2022), the ES yield per egg is 11.04% (w/w), while the ESM yield per egg is 0.47% (w/w), with ESM recovery from ES estimated at around 4% (w/w). Supporting the previous statement, Jain and Anal (2016) stated that both ES and

ESM constitute 11% of the total egg weight, whereas the ESM, accounts for approximately 5% of the total ES weight. Additionally, ES microstructural characteristics undergo modifications based on various physiological conditions such as age, diet, and disease. Notably, aging hens exhibit significant changes in ES microstructure, accompanied by a marked reduction in its mechanical properties (Rodriguez-Navarro et al. 2007).

The longitudinal section depicting the interior contents of a chicken egg can be observed in Fig. 1. The egg comprises of the calcified ES, albumen (egg white), and membranes surrounding the yolk. The location of ESM in the egg is represented by the blue lines, where the ESM is further described by distinguishing its three substructures: the outer ESM, the inner ESM, and the limiting membrane (Shi et al. 2021).

The images of scanning electron microscopy (SEM) show distinct morphologies and thicknesses for the outer ESM and the inner ESM, where the outer ESM has a thickness of 50 to 70 μm, and the inner ESM has a thickness of 15 to 26 μm. The outer ESM, which has fiber diameters ranging from 1 to 7 μm, is attached to the eggshell. On the other hand, the inner ESM has fiber diameters that are smaller, measuring from 0.1 to 3 μm (Zhou et al. 2011). These membranes serve to retain albumen and act as a barrier against bacterial penetration (Nakano et al. 2003). Notably, a smooth layer of amorphous organic colloid, known as the "limiting film", is present on the inner face of the fiber membrane that comes into contact with the egg white. This film serves to prevent egg white leakage (Zhou et al. 2011). Located in between the ES and the albumen, the ESM provides a non-mineralized structure to assist the mineralization of ES on the outside and inhibits the albumen mineralization on the inside for ES formation due to its biopolymeric fibrous net characteristic (Park & Kim 2017). ESM also provides a biological homeostatic environment in which the developing embryo in fertilized eggs can thrive (Rose-Martel et al. 2015),

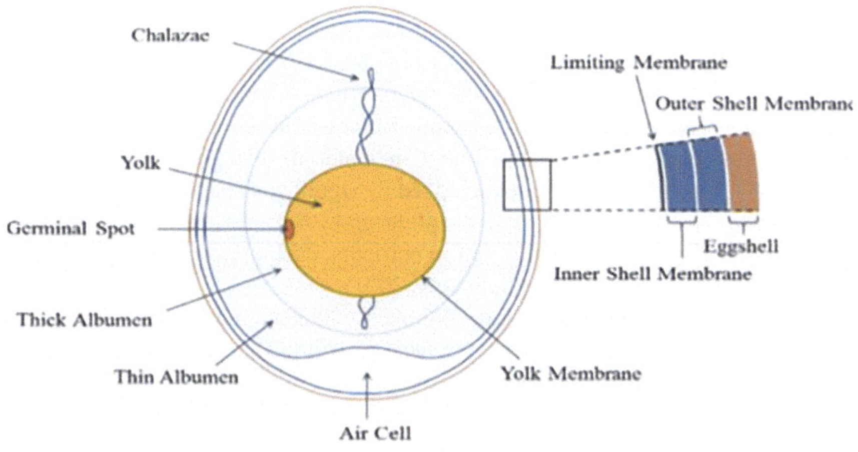

Fig. 1 Interior contents of a chicken egg (Shi et al. 2021)

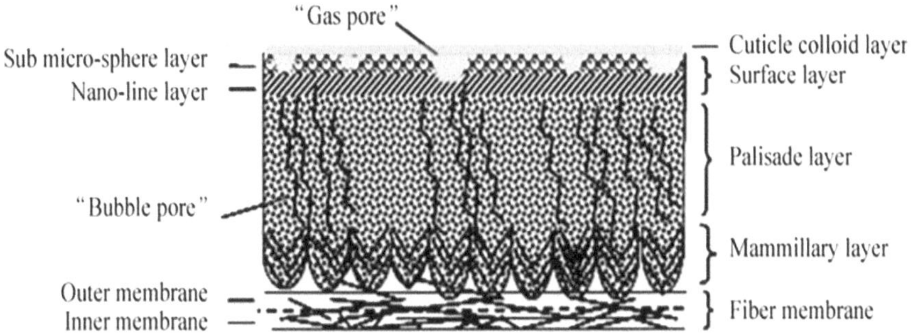

Fig. 2 Cross-sectional view of chicken eggshell (Zhou et al. 2011)

The cross-sectional view of chicken ES is depicted in Fig. 2. Extending from the exterior to the egg white, the ES exhibits anisotropic structures in the radial direction, primarily comprising the surface layer, palisade layer, and mammillary layer. This shell can be outlined into these layers, with a particular colloid layer overlaying the shell and a keratin-fiber membrane beneath it. The surface layer showcases $CaCO_3$ morphologies characterized by nano-rods covered with submicro spheres, the palisade layer reveals a rhombohedral-cleavage pattern, and the mammillary layer features slices. Conventional "gas pores" are present in the surface layer while "bubble pores" are found in the palisade and mammillary layers (Zhou et al. 2011). As a result, the ESM contains a large surface area of stable, water-insoluble fibers arranged in a complex lattice network (Tsai et al. 2006).

2.2 *Chemical Properties of ESM*

The ESM is composed of 69.2% protein, 2.7% fat, 1.5% moisture, and 27.2% ash (King'Ori 2011) with a 1.358 g/cm^3 density (Mittal et al. 2016). There are 472 distinct types of proteins found in the 90% protein fraction (Han et al. 2023). Enzymatic hydrolysis of ESM is challenging due to the presence of robust structural proteins, particularly keratin, found in materials like feathers, hair (Zhao et al. 2019). The chicken ESM, characterized by significant stability through extensive desmosine, isodesmosine, and disulfide cross-linkages, is highly insoluble (Kulshreshtha et al. 2022). ESM is rich in nutrients, encompassing fibrous proteins and glycosaminoglycans (Hisamuddin et al. 2021).

Structurally, ESM comprises of cross-linked collagens, numbered I, V, and X, glycosaminoglycans, egg white proteins such as ovotransferrin and lysozyme, and eggshell matrix proteins, for instance, ovocalyxin-36 (Shi et al. 2021). Dominantly proteinaceous, proteomics and bioinformatics analysis have identified over 500 proteins in ESM, with noteworthy bioactive proteins including ovocalyxin-36, lysozyme, and avian β defensins (AvBDs) (Ahmed et al. 2017). Protein constitutes the major component in ESM, reported as 73 g/100 g with minimal fat content (0.3 g/

Table 1 Amino acid composition in inner and outer ESM (Nakano et al. 2003)

Amino Acid	Inner eggshell membrane (mol%)	Outer eggshell membrane (mol%)
Proline	11.6 ± 0.7	12.0 ± 0.9
Glutamic acid	11.1 ± 0.4	11.9 ± 0.3
Glycine	11.1 ± 0.2	10.6 ± 0.2
Serine	9.2 ± 0.2	9.2 ± 0.0
Aspartic acid	8.4 ± 0.4	8.8 ± 0.1
Valine	7.2 ± 0.2	7.9 ± 0.1
Threonine	6.9 ± 0.0	6.9 ± 0.2
Arginine	5.7 ± 0.3	5.8 ± 0.2
Leucine	5.6 ± 0.5	4.8 ± 0.2
Alanine	4.6 ± 0.2	4.1 ± 0.2
Histidine	4.1 ± 0.4	4.3 ± 0.4
Lysine	3.6 ± 0.2	3.4 ± 0.2
Isoleucine	3.3 ± 0.4	3.4 ± 0.1
Methionine	2.3 ± 1.0	2.3 ± 1.0
Tyrosine	2.2 ± 0.1	1.7 ± 0.3
Phenylalanine	1.6 ± 0.1	1.5 ± 0.1
Hydroxyproline	1.5 ± 0.3	1.4 ± 0.4

100 g). Additionally, ESM comprises 10.7 g/100 g of moisture content and an ash content of 6.4 g/100 g (Jain & Anal 2016).

In fact, the approximate 500 proteins found in both ES and ESM, are four to five times greater than other egg components such as egg yolk (316 proteins), albumen (148 proteins), and vitelline membrane (137 proteins) (Baláž 2014). According to Zhao et al. (2019), any presence of proline, histidine, tyrosine, valine, or leucine enhances the antioxidant ability of a peptide. The antioxidant potential of ESM is evident with the abundance of proline which is depicted in Table 1, which showed the amino acid composition, in inner and outer ESM.

Usually, bioactive peptides exhibit short amino acid sequences, typically comprising around 2 to 20 residues, embedded within their parent protein. While these peptides possess functional effects, their release from the complete protein sequence requires processes such as enzymatic hydrolysis, microbial fermentation, or solvent extraction (Borrajo et al. 2020).

2.3 Characterization of ESM

Mechanical behavior of ESM has also been studied. At low strain levels, the mechanical behavior of ESM replicates other biopolymer materials such as collagen as observed by its non-linear elasticity due to the entropic and enthalpic mechanisms of the molecules. Torres et al. (2010) reported that, contradictory, at high strain levels, the ESM has a mechanical behavior similar to other fiber networks. Due to the presence of protein fiber, ESM is characterized by its low thermal stability and its thermal decomposition begins at around 55 °C where collagen undergoes thermal denaturation (Torres et al. 2013). The peak of denaturation temperature can be increased by increasing the amount of water in the ESM since the H_2O molecules will form hydrogen bonds with ESM amide groups. Water can also act as a plastifying agent for ESM.

Additionally, there is a study by Oda et al. (1998), extracted from Baláž (2014) discussing the ES and ESM resistance to mechanical penetration. This can be inferred by the ESM and ES being held by a strong adhesive agent of the ESM. Higher penetration energy outside the ES is also required to produce energy that will break the membrane and peel the ES.

3 ESM Applications

3.1 ESM for Various Applications

Park and Kim (2017) laid out several advantages of ESM that spark the interest of the industries. As an abundant agri-industrial waste, ESM is considered cheap and readily available. It is also non-toxic and environmentally friendly. Most importantly, it consists of many natural proteins that could be readily modified to suit various purposes of the industries. Hence, following this, a vast number of ESM applications have been identified such as biosensors, nanoparticles bio-templates, electrochemistry as a separator in the supercapacitor and its role as absorbent. This also includes its various practice in the biomedical field due to its biocompatibility, including tissue and bone regeneration, therapeutics for joint, and wound healing for both in vitro and in vivo applications (Baláž 2014). Additionally, the hydrolyzed ESM had been proven to exhibit antimicrobial, anti-inflammatory, anti-wrinkle, and antioxidant activities (Yoo et al. 2014) with various other applications in multiple biochemical fields since it is discovered that hydrolyzing ESM will help improve and enhance the presence of its bioactive peptides (Zhao et al. 2019).

Clinical studies have demonstrated the efficiency of ESM in alleviating joint inflammation and promoting wound healing. Hydrolysates derived from ESM exhibit activities such as free radical scavenging, DNA protection, skin whitening, and defense against wrinkling (Zhao et al. 2019). The remarkable bioactivity and favorable safety profile of natural antioxidant peptides make them promising candidates for mitigating oxidative stress, surpassing synthetic antioxidants and drugs (Zhu et al. 2022). The biological functions of ESM are various and include antimicrobial, anti-inflammatory, anti-wrinkle, and antioxidant properties (Kulshreshtha et al. 2022). Advantages of ESM include its cost-effectiveness, easy to obtain from the abundant industrial and household waste, environmental friendliness, and non-toxic nature. Its composition of natural proteins, rich in amino acids, facilitates surface functionalization and adsorption. Additionally, ESM's adaptability to modify through processes enhances its versatility, complementing its chemical characteristics (Park et al. 2016).

3.2 ESM as Antioxidant

Exposure to a variety of external stressors such as cigarette smoke, ionizing radiation, air pollution, UV light, heavy metal ions, and hydrogen peroxide makes living organisms more vulnerable to oxidative stress. Antioxidant activity is a type of bioactivity exhibited in peptides due to its role in preventing oxidative stress and repairing the biological damage of free radicals and reactive oxygen species (ROS) (Santana et al. 2016).

According to Lee et al. (2017), peptides prevent oxidative damage to biological molecules through a variety of mechanisms including ROS inactivation, metal ion chelation, and free radical scavenging. ESM hydrolysates produced with a range of alkaline proteases show good radical scavenging action due to their collagen composition (Shi et al. 2014). Huang et al. (2010) in their article reviewed that ESM is comprised of several amino acids that are verified to exhibit antioxidant activities such as arginine, lysine, histidine, tryptophan, cysteine, leucine, valine, proline, glutamic acid, and β-hydroxyl tryptophan.

3.3 ESM as Antimicrobial Agent

Antimicrobial peptides (AMPs) offer a wide spectrum of inhibitory activities against bacteria, viruses, parasites, and fungi. The proliferation of antibiotic-resistant microbes, as well as growing concerns about the usage of antibiotics culminated the development of AMPs from a variety of sources including proteins (Huan et al. 2020).

The antibacterial activity of ESM had been recorded in multiple works. Poland and Sheldon (2001) studied the antibacterial activity of ESM against numerous kinds of bacteria. They demonstrated that ESM could potentially be used to modify the heat-resistant activity of the bacteria such as *E coli.* Other than that, Mine et al., (2003) had also proven how ESM had successfully been able to interact and affect the bacterial membrane integrity.

3.4 ESM as Anti-Inflammatory Agent

In living organisms, cytokines act as regulators of the responses of the host against infection, inflammation, and immune responses. Certain cytokines are pro-inflammatory and certain cytokines are anti-inflammatory which help in the treatments to reduce inflammation and foster wound healing.

To validate the role of ESM as an anti-inflammatory agent, the study of Devore and Long (2013) demonstrated the reduction of plasma pro-inflammatory cytokines which correspond to inflammation and pain following oral treatment of cells with ESM hydrolysate. Additionally, a study by Yoo et al. (2014) displayed that the ESM hydrolysate had been successful in hindering the initiation of cell inflammation and suppressing skin inflammation. The ESM is said to contain both anti-lipopolysaccharide and anti-IFN-γ-induced inflammation activities to protect the skin from inflammation.

3.5 ESM as Anti-aging Agent

There are active researches on developing ESM as raw material for improving skin function, to be consumed as a functional food to support skin health. Sim et al., (2023) reported that feeding ESM was found to improve skin moisture and reduce wrinkles on hairless mouse model. Their results also showed that oral administration of ESM restored moisture in the dorsal skin tissue of mice, reversed the increased trans-epidermal water loss and reduction of mRNA expression of hyaluronic synthases induced by UVB irradiation. Furthermore, UVB irradiation-induced collagen degradation was inhibited, and the expression of the collagenase MMP was also reduced. They also reported that, the mRNA expression of the antioxidant enzyme superoxide dismutase 1 (SOD1), which was reduced on UVB irradiation, was restored on ingestion of the ESM.

Other researches have also showed ESM to have a beneficial effect as nutricosmetic ingredients, whereby, Yoo et al. (2014) reported that ESM exhibited anti-wrinkling and anti-inflammatory activities in vitro. Additionally, Kalman and Hewlings (2020) reported that in their clinical study, oral administration of ESM has beneficial effects on the skin, hair, and nails. Ruff et al. (2009) in their clinical trials have also indicated the improvement of connective tissue disorders on treatment with ESM. All these results confirmed that oral ingestion of the ESM is beneficial and revealed the potential of ESM as a nutricosmetic material with anti-wrinkle and skin moisturizing properties. However, the mechanism underlying the action of ESM contributing to anti-aging remains unclear.

4 Recovery of Bioactive Proteins from ESM

4.1 Separation of ESM from ES

To recover bioactive peptides from ESM, the first step is to separate ESM from ES, which can be achieved through either mechanical or chemical methods. Mechanical separation involves manually peeling off the ESM from the ES. Baláž (2014) highlighted that this mechanical approach can only separate the inner ESM and limiting membrane (LM) because the outer ESM is firmly embedded in the ES. Despite this limitation, it is considered a favorable extraction process due to its simplicity and cost-efficiency, requiring no chemical intervention.

Alternatively, chemical treatment can be employed for the separation of ESM from ES. This involves the use of chemical such as hydrochloric acid, ethylenediamine tetra-acetic acid (EDTA), or dilute acetic acid. In this method, the ES is immersed in the acid to facilitate the separation from the ESM. Park and Kim (2017) suggested that ES can be incubated for 18 h in 100% acetic acid at 4 °C as a part of the chemical separation. The ESM can later be washed with deionized water and stored in a 5 mM potassium phosphate buffer (pH 7) at 4 °C. While acid treatment effectively removes contaminating shell minerals, it may alter the chemical composition of the membranes, which is not preferred (Kulshreshtha et al. 2022).

Other methods for separation of ESM from ES include non-chemical approaches like water-based flotation or air turbine-based separation, alongside microwave-assisted separation. Additional methods include vacuum treatment, cyclone, particle size reduction, cavitation, processing chamber impact zone, mechanical abrasion, liquid agitation, and differential flotation (Kulshreshtha et al. 2022).

4.2 Enhancing ESM Extraction by Ultrasonication

One of the obstacles to profit from the use of bioactive peptides in ESM is their limited solubility due to the cross-linking of disulfide bonds. The challenge of insolubility in ESM poses a hindrance to its research progress and utilization in relevant studies. Jain and Anal (2016) suggested to include the ultrasonication technique to enhance the extraction of ESM. They reported an amplitude of 95.74%, a duration of 28.06 min, and a solid-to-solvent ratio of 1:30 (g:mL), to be the optimal ultrasonication condition for maximal ESM extraction. When compared to the control samples (untreated enzymatic hydrolysates), the hydrolysates with ultrasonic pretreatment had higher protein concentrations and higher degrees of hydrolysis, which improved their functional characteristics. Because bubbles are created and cavitational effects occur, the ultrasonication process generates shearing forces and heating effects.

The macrostructure of ESM relies on proteins interconnected through lysine-derived cross-links, with collagen constituting approximately 10% of the total protein content. Ultrasound energy has the capacity to induce changes in matter, impacting the molecular structure of proteins, facilitating the extraction of lipids and proteins, modifying enzyme activity, or enhancing the functional properties of certain proteins. One mechanism, the ultrasonication fragmentation effect, leads to a reduction in the particle size of the sample. Studies involving vegetable material suggest that ultrasonication improves hydration and fragmentation of the tested samples (Toma et al. 2001). In this context, ultrasonication may facilitate the incorporation of water into the sample through channels and pores, resulting in faster swelling. Consistent with this, ultrasonication could aid in the fragmentation of ESM, simultaneously promoting its swelling with the reaction medium solution, thereby increasing its solubilization rate (Marcet et al. 2018). This study presented as the first time that ultrasonication is used as an extraction method of ESM but more importantly, it demonstrated that there is no significant impact on the composition of the solubilized protein, affirming that ultrasonication can improve protein extraction from the ESM without altering the final product obtained.

4.3 Solubilization of ESM into SEP

Despite the ESM's natural unique structure which can be taken advantage of in various industries, some characteristics of ESM present difficulties in the application of its natural form. This issue can be solved through the modification of the morphology and the composition of ESM. Among the procedures available include thiolation, pore size modification, mineralization, and the transformation of ESM to soluble ESM protein (SEP) (Baláž 2014).

For example, due to its insoluble properties from its cross-linked disulfide bond (Yi et al. 2003), ESM which is located in between the ES and the albumen was further modified chemically into SEP to strengthen its mechanical properties and ensure easier transformation of its shape and size. To allow easy control of ESM into its desired size and shape, Yi et al. (2003) determined the optimal conditions of conversion to SEP by heating ESM at 90 °C for 12 h. 3-mercaptopropionic acid can be used as reducing agent, however, since 3-mercaptopropionic acid is considered toxic to human health (Shi et al. 2014), many other reducing agents can be used including 2-mercaptoethanol, dithiothreitol (DTT), and sodium metabisulfite. Another research by Marcet et al. (2018) used alkali solution treatment on ESM, resulting in the attainment of SEP, constituting close to 100% of the total ESM. Following the solubilization process, the mixture then undergoes centrifugation to eliminate any traces of insoluble solid (Yi et al. 2003).

4.4 Hydrolysis of SEP of ESM

Hydrolysis is the process of protein digestion that is done by targeting specific peptide cleavage bonds to break down proteins into amino acids and peptides of different sizes (Tapal & Tiku 2019). There are various factors affecting the degree of hydrolysis including the type of enzyme, the enzyme dosage, the duration of hydrolysis, and the hydrolysis condition such as the temperature and pH. Chemical and enzymatic hydrolysis are both conventional approaches for producing protein hydrolysates and peptides. The enzymatic method is typically preferred due to its high substrate specificity and mild reaction conditions (Jain & Anal 2016). These hydrolysis processes aim to break down the polypeptide bonds in the ESM protein structure, yielding amino acids and peptides (Hisamuddin et al. 2021). Various proteases, including alcalase, trypsin (Kadir et al. 2022), alkaline protease, pepsin, chymotrypsin, neutral protease, and papain have been employed to generate hydrolysates with enhanced antioxidant activity (Zhao et al. 2019). Notably, alcalase demonstrated high efficacy (Zhu et al. 2022). In a comparative study, three proteases, namely, alcalase, bromelain, and papain, were used for enzymatic hydrolysis over two different durations, four and eight hours. The hydrolysates that were obtained after eight hours with alcalase demonstrated enhanced antioxidant characteristics, whereas those that were obtained after four hours with bromelain demonstrated increased antimicrobial efficacy. Another study by Borrajo et al. (2020) used common enzymes such as alcalase, papain, pepsin, and trypsin to hydrolyze protein from porcine liver.

Sodium sulfite (Na_2SO_3) is capable of reducing disulfide bonds and has found applications in food processing, as well as in the extraction of keratin from feathers. Similarly, alkaline proteases prove effective in extracting polypeptides from both animal and plant proteins, as demonstrated in processes like extracting antioxidant peptides from sorghum proteins. Consequently, both Na_2SO_3 and alkaline protease show potential for use in the hydrolysis of ESM (Zhao et al. 2019).

5 Purification of Bioactive Proteins

5.1 Purification by Aqueous Two-Phase System (ATPS)

The purification process is the most important stage of production, making up more than 70% of the value of downstream processing. Recently, a technique known as aqueous two-phase systems (ATPS) is utilized as an alternative for the purification of biomolecules due to its advantages of high productivity, simple operation, short processing time, cost-effective and easy to up-scale (Tan et al. 2014). It is a liquid–liquid extraction method used to separate biomolecules and purify peptides by the mixture of two immiscible liquids. ATPS is known to be a simple and mild process that is free from protein denaturation due to its high-water content. It also has a low interfacial tension and stabilizing effect supplied by the polymers to maintain the protein's native structures (Asenjo & Andrews 2012).

In essence, ATPS occurs when two immiscible water-soluble liquids mix above their concentration until two separate immiscible phases are formed at equilibrium. (Iqbal et al. 2016). The separation result then allows the distribution of the biological substances between the phases. Without any organic solvent addition, ATPS is very much favorable, especially in the food and pharmaceutical industries to extracting biomolecules of low molecular weight (Dumas et al. 2020). Generally, after partitioning, the bioactive peptides will be accumulated in the top phase due to their hydrophobicity and less polarity toward the polymer while other soluble and particulate matter will be at the bottom phase (Yusof 2015). According to Iqbal et al. (2016), Albertsson's model (1987) suggested a few driving forces that determine the partition behavior of the two liquid phases such as electrochemical, hydrophobicity, bio-specific affinity, size, and conformation-dependent where hydrophobicity is considered the most dominant. The single-step operation had been proven to be cost, resource, and time-efficient while minimizing the denaturation risk and loss of target protein.

Before conducting ATPS, a phase diagram should be designed to reflect the concentration of top and bottom components necessary to produce the two-phase system, including the volume ratio (Raja et al. 2011). The phase diagram in Fig. 3 shows the binodal curve (TCB line) that separates the above curve region (two aqueous phase regions) and the below curve region (one phase). Methods of turbidimetric titration, cloud-point titration, and node determination method can be used to construct the binodal curve (Hatti-Kaul 2000). The binodal curve is constructed by connecting the points of the end node where each node reflects the final composition of PEG/salt before it becomes one phase. To do so, the correlations between the two compositions should be understood.

Apart from the binodal curve, the phase diagram also consists of a tie-line (TB line) as illustrated in Fig. 3. The tie-line is a straight line that connects two endpoints on the binodal curves that reflect the top and bottom phases' final concentration (Pereira & Coutinho 2020). All the nodes along each tie-line (S1, S2, S3) have equal top and bottom phase equilibrium composition. Another crucial parameter is the tie-line length (TLL) which is calculated using the Pythagoras theorem to evaluate the system composition consequences on partitioning components (Hatti-Kaul 2000).

Meanwhile, the critical point reflects the uniform volume of the top and bottom phase or V_R of 1.00 (equal volume ratio of top and bottom phase) and TLL of 0. The critical point can be calculated by extrapolating the tie-line until it approaches the binodal curve. Since tie lines are commonly parallel, they could be easily extrapolated based on the slope (Yusof 2015). The vertical axis represents the top phase composition, and the horizontal axis represents the bottom phase composition. The curve represents the division of both phases. Additionally, the partition coefficient can be determined based on the ratio of the top phase equilibrium concentration and the bottom phase equilibrium concentration. ATPS can also be characterized by its purification factor, which is the ratio of the product concentration in any phase of your preference to its initial concentration (Raja et al. 2011).

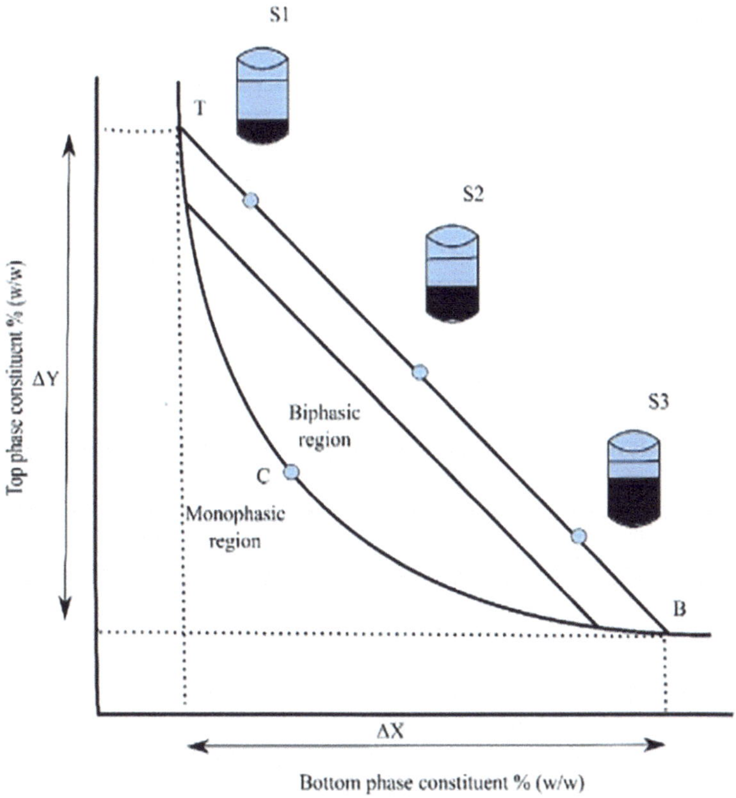

Fig. 3 Phase Diagram of ATPS (TCB = Binodal curve, TB = Tie line, C = critical point, T = composition of the top phase, B = composition of the bottom phase, and S1, S2, and S3 = total composition of ATPS phases.) (Iqbal et al. 2016)

5.2 Purification by Chromatography Method

Alternatively, while ATPS is quite a new method for protein purification, chromatography has been used conventionally in purifying bioactive proteins from different sources. Based on the properties of the proteins, such as their sizes, charges, and hydrophobicity, as well as polarity in which the molecules are partitioned between the stationary and mobile phases, column chromatography can also be used to separate and purify any protein of interest (Yusof 2011 2015).

According to Zhao et al. (2019), ion exchange chromatography is utilized frequently in the sectors of food and fermentation. It has been proven to be beneficial due to its high affinity, rapid separation recovery, and mild reaction conditions. Among all, the most popular method for separating proteins would be cation exchange chromatography. For example, tilapia frame protein had been previously purified using ion exchange chromatography, gel filtration chromatography, and reversed-phase high-performance liquid chromatography (RP-HPLC) (Fan et al. 2012).

Despite that, chromatography could not work independently without first precipitating the protein, resulting in a very tedious and time-consuming process. As multiple unit operation is sometimes involved during the process, a lower yield of the product might be recovered due to the increase in the loss of target molecules.

6 Analysis of Peptides Hydrolysates

6.1 Degree of Hydrolysis

Upon the hydrolysis of protein, it is pertinent to have the degree of hydrolysis determined. The degree of hydrolysis, defined as the percentage of peptide bonds cleaved, can be calculated by the determination of free amino groups by reaction with TNBS according to Adler-Nissen (1979) with slight modifications (Megías et al. 2007). The total number of amino groups will be determined in a sample that was 100% hydrolyzed at 110 °C for 24 h in 6 N HCl (10 mg sample in 4 mL HCl).

6.2 Protein Quantitative Analysis

The protein concentration is quantitatively determined by Bradford (1976) protein analysis method using UV–VIS spectrophotometer at the wavelength of 595 nm. Bovine serum albumin (BSA) is used as a standard for protein assay, where the absorbance of seven concentrations of BSA, 2.0, 1.5, 1.0, 0.75, 0.5, 0.25 and 0.125 mg/ml are measured. The plotted standard curve of absorbance at 595 nm against the concentration of BSA will be used as a guide to determine the concentration of protein in the samples.

For the sample preparation, 5 μL of the sample and 250 μL of Bradford dye reagent are pipetted into a clean 96-well microplate. The mixture is repeatedly mixed in the wells. The sample is then incubated at room temperature for between 5 min and 1 h. Using the UV–VIS spectrophotometer, the absorbance at 595 nm is measured after about 10 min of incubation.

6.3 Protein Qualitative Analysis

Sodium Dodecyl Sulfate–Polyacrylamide Gel Electrophoresis (SDS-PAGE) can be conducted to qualitatively analyze the proteins present in the ESM sample based on their molecular weight. To cast the gel, the resolving gel and the stacking gel of

Table 2 Composition of SDS-PAGE casting gel

Type of gel	Gel %	Sterile dH2O (mL)	30% Acrylamide (mL)	Gel buffer (mL)	10% SDS (uL)	10% APS (uL)	TEMED (uL)
Resolving gel	12	1.7	2	1.25	50	50	5
Stacking gel	4	3.05	0.65	1.25	50	50	10

Table 2 were poured into the assembled apparatus and were allowed to stand until each gel is solidified. The comb was inserted into the top of the spacers.

For the sample preparation, samples will be added to the Laemmli sample buffer at a 1:1 ratio and heating them for 5 min at 95 °C. Before loading the samples, the running buffer will be filled into the electrophoresis chamber. A protein ladder containing a wide range of MW markers is loaded into the first well using a micropipette, followed by all the samples. Once sample loading is done, electrophoresis will be carried out at 120 V for 45 min, or until the dye front moves to the bottom of the casting gel. After the electrophoresis, the protein in the gel will undergo fixing, staining, and destaining. After destaining, the gels are stored in a storage solution and is photographed as required to observe the resolved proteins.

6.4 Anti-Oxidative Activity Assays

The anti-oxidative properties of the hydrolysates can be tested using any of the two methods of radical scavenging assays:

1. The first method is termed "DPPH antioxidant activity assay". To prepare the DPPH working solution (Preito 2012), the DPPH powder was dissolved with 10 ml of 99.5% ethanol. The DPPH working solution was wrapped in foil and kept in the fridge to reduce degradation for a week. To conduct the experiment, 200 uL of the mixture was added to each 96-well plate consisting of 100 uL of DPPH solution, 80 uL of assay buffer, and 20 uL of the sample. Blank 1 consisted of 100 uL of DPPH solution, 80 uL of assay buffer, and 20 uL of the PBS solvent while Blank 2 consisted of 80 uL of assay buffer and 100 uL of ethanol, and 20 uL of the PBS solvent. All measurements were conducted in triplicate. The 96-well microplate reader was then incubated in the dark for 30 min at room temperature and using a spectrophotometer, the absorbance of the sample was measured at 517 nm. According to Xiao et al. (2020), the inhibition ratio, or the scavenging activity of the sample was measured by:

$$\text{Inhibition}\,\% = (\text{A1} - \text{A2}) \times 100/\text{A1} \tag{1}$$

where:

A1 = Absorbance of Blank 1–Blank 2

A2 = Absorbance of Sample–Blank .

2. The second method of radical scavenging assay is by conducting the "Ferric Reducing Antioxidant Power (FRAP)" assay, whereby, the antioxidant capacity of the can be estimated spectrophotometrically following the procedure of Benzie and Strain (1996). The method is based on the reduction of Fe^{3+} TPTZ complex (colorless complex) to Fe^{2+}-tripyridyltriazine (blue colored complex) formed by the action of electron donating antioxidants at low pH. This reaction is monitored by measuring the change in absorbance at 593 nm. The ferric reducing antioxidant power (FRAP) reagent was prepared by mixing 300 mM acetate buffer, 10 ml TPTZ in 40 mM HCl and 20 mM $FeCl_3.6H_2O$ in the proportion of 10:1:1 at 37 °C.

Freshly prepared working FRAP reagent was pipetted using 1–5 ml variable micropipette (3.995 ml) and mixed with 5 μl of the appropriately diluted plant sample and mixed thoroughly. An intense blue color complex was formed when ferric tripyridyl triazine (Fe^{3+}-TPTZ) complex was reduced to ferrous (Fe^{2+}) form and the absorbance at 593 nm was recorded against a reagent blank (3.995 ml FRAP reagent + 5 μl distilled water) after 30 min incubation at 37 °C. All the determinations were performed in triplicates. The calibration curve was prepared by plotting the absorbance at 593 nm *versus* different concentrations of $FeSO_4$. The concentrations of $FeSO_4$ were in turn plotted against the concentration of standard antioxidant (6-hydroxy-2,5,7,8-tetramethylchroman-2-carboxylic acid or Trolox. The FRAP values were obtained by comparing the absorbance change in the test mixture with those obtained from increasing concentrations of Fe^{3+} and expressed as mg of Trolox equivalent per gram of sample.

7 Conclusion

ESM is considered a valuable and promising raw material to enhance the circular economy for the production of natural bioactive peptides. Due to their abundance and valuable compositions, this chapter presented a part of the solution to solve the issue of the waste management of agri-industrial wasteful eggshell membranes (ESM) as well as to optimize the valuable resource utilization. This overview of ESM functional characteristics, applications, and recovery processes could serve as the foundation for pilot studies to establish the feasibility of the process to be scaled up for industrial applications. With that, the recovery of bioactive proteins from ESM is highly encouraged using various alternatives for a variety of purposes within the nutraceuticals and pharmaceutical industry.

References

Adler-Nissen, J. (1979). Determination of the degree of hydrolysis of food protein hydrolyzates by trinitrobenzenesulfonic acid. *Journal of Agriculture and Food Chemistry, 27*, 1256–1262.

Ahmed, T. A. E., Henri-Pierre, S., & Hincke, M. T. (2017). In-depth comparative analysis of the chicken eggshell membrane proteome. *Journal of Proteomics, 155*, 49–62. https://doi.org/10.1016/j.jprot.2017.01.002

Albertsson, P. Å., Cajarville, A., Brooks, D. E., & Tjerneld, F. (1987). Partition of proteins in aqueous polymer two-phase systems and the effect of molecular weight of the polymer. *Biochimica et Biophysica Acta (BBA)*—General Subjects, *926*, 87–93. https://doi.org/10.1016/0304-4165(87)90185-1

Asenjo, J. A., & Andrews, B. A. (2012). Aqueous two-phase systems for protein separation: Phase separation and applications. *Journal of Chromatography A, 1238*, 1–10. https://doi.org/10.1016/j.chroma.2012.03.049

Baláž, M. (2014). Eggshell membrane biomaterial as a platform for applications in materials science. *Acta Biomaterialia, 10*(9), 3827–3843. https://doi.org/10.1016/j.actbio.2014.03.020

Benzie, I. F., & Strain, J. J. (1996). The ferric reducing ability of plasma (FRAP) as measurement of "antioxidant power" The FRAP assay. *Analytical Biochemistry, 239*, 70–76.

Borrajo, P., Pateiro, M., Gagaoua, M., Franco, D., Zhang, W., & Lorenzo, J. M. (2020). Evaluation of the antioxidant and antimicrobial activities of porcine liver protein hydrolysates obtained using alcalase, bromelain, and papain. *Applied Sciences, 10*(7), 2290. https://doi.org/10.3390/app10072290

Bradford, M. M. (1976). A rapid and sensitive method for the quantitation of microgram quantities of protein utilizing the principle of protein-dye binding. *Analytical Biochemistry, 72*, 248–254. https://doi.org/10.1016/0003-2697(76)90527-3

Devore, D. P., & Long, F. D. (2013). Anti-inflammatory activity of eggshell membrane and processed eggshell membrane preparations. *U.S. Patent No. 8,580,315*. U.S. Patent and Trademark Office.

Dumas, F., Roger, E., Rodriguez, J., Benyahia, L., & Benoit, J. P. (2020). Aqueous Two-Phase Systems: Simple one-step process formulation and phase diagram for characterisation. *Colloid and Polymer Science, 298*(12), 1629–1636. https://doi.org/10.1007/s00396-020-04748-8

Fan, J., He, J., Zhuang, Y., & Sun, L. (2012). Purification and identification of antioxidant peptides from enzymatic hydrolysates of tilapia (*Oreochromis niloticus*) frame protein. *Molecules, 17*(11), 12836–12850. https://doi.org/10.3390/molecules171112836

Han, C., Chen, Y., Shi, L., Chen, H., Li, L., Ning, Z., Zeng, D., & Wang, D. (2023). Advances in eggshell membrane separation and solubilization technologies. In *Frontiers in Veterinary Science, 10*.https://doi.org/10.3389/fvets.2023.1116126

Hatti-Kaul, R., ed. (2000) Aqueous two-phase systems. methods Biotechnol, *11*, *Methods and Protocols*, Humana Press.

Hincke, T. H., Nys, Y., Gautron, J., Mann, K., Rodriguez-Navarro, A. M., & McKee, M. (2012). The eggshell: Structure, composition and mineralization. *Frontiers in Bioscience, 17*, 1266–1280. https://www.researchgate.net/publication/51895246

Hisamuddin, N. S. N., Muttalib, S. A. A., Kadir, N. F. S. A., Mohamad, S. A. S., Ahmad, N. M., & Tan, E. T. T. (2021). Laboratory scale eggshell membrane separation and hydrolysis. *Symposium of Research and Innovation for Food*, 104–107,https://doi.org/10.1007/s00604-014-1225-6

Huan, Y., Kong, Q., Mou, H., & Yi, H. (2020). Antimicrobial peptides: Classification, design, application and research progress in multiple fields. *Frontiers in Microbiology, 11*(582779), 1–21. https://doi.org/10.3389/fmicb.2020.582779

Huang, X., Zhou, Y., Ma, M., Cai, Z., & Li, T. (2010). Chemiluminescence evaluation of antioxidant activity and prevention of DNA damage effect of peptides isolated from soluble eggshell membrane protein hydrolysate. *Journal of Agricultural and Food Chemistry, 58*(23), 12137–12142. https://doi.org/10.1021/jf101728d

Iqbal, M., Tao, Y., Xie, S., Zhu, Y., Chen, D., Wang, X., Huang, L., Peng, D., Sattar, A., Abu, M., Shabbir, B., Hussain, H. I., Ahmed, S., & Yuan, Z. (2016). Aqueous two-phase system (ATPS):

An overview and advances in its applications. *Biological Procedures Online*, 1–18. https://doi.org/10.1186/s12575-016-0048-8

Jain, S., & Anal, A. K. (2016). Optimization of extraction of functional protein hydrolysates from chicken eggshell membrane (ESM) by ultrasonic assisted extraction (UAE) and enzymatic hydrolysis. *LWT-Food Science and Technology, 69*, 295–302. https://doi.org/10.1016/j.lwt.2016.01.057

Kadir, N. F. S. A., Ahmad, N. M., Mohamad, S. A. S., Chua, L. S., & Tan, E. T. T. (2022). Effect of enzymatic hydrolysis on the structural and protein solubility of eggshell membrane. *Jurnal Teknologi, 84*(6), 81–88. https://doi.org/10.11113/jurnalteknologi.v84.18654

Kalman, D. S., & Hewlings, S. (2020). The effect of oral hydrolyzed eggshell membrane on the appearance of hair, skin, and nails in healthy middle-aged adults: A randomized double-blind placebo-controlled clinical trial. *Journal Cosmetic Dermatology, 19*, 1463–1472. https://doi.org/10.1111/jocd.13275

King'ori, A. M. (2011). A review of the uses of poultry eggshells and shell membranes. *International Journal of Poultry Science, 10*(11), 908–912, ISSN 1682–8356 © Asian Network for Scientific Information.

Kulshreshtha, G., Diep, T., Hudson, H. A., & Hincke, M. T. (2022). High value applications and current commercial market for eggshell membranes and derived bioactives. In *Food Chemistry, 382*. Elsevier Ltd. https://doi.org/10.1016/j.foodchem.2022.132270

Lee, J. H., Moon, S. H., Kim, H. S., Park, E., Ahn, D. U., & Paik, H. D. (2017). Antioxidant and anticancer effects of functional peptides from ovotransferrin hydrolysates. *Journal of the Science of Food and Agriculture, 97*, 4857–4864. https://doi.org/10.1002/jsfa.8356

Marcet, I., Salvadores, M., Rendueles, M., & Díaz, M. (2018). The effect of ultrasound on the alkali extraction of proteins from eggshell membranes. *Journal of the Science of Food and Agriculture, 98*(5), 1765–1772. https://doi.org/10.1002/jsfa.8651

Megías, C., Pedroche, J., Yust, M., Girón-Calle, J., Alaiz, M., Milan, F., & Vioque, J. (2007). Affinity purification of copper-chelating peptides from sunflower protein hydrolysates. *Journal of Agriculture and Food Chemistry, 55*(16), 6509–6514. https://doi.org/10.1021/jf0712705

Mine, Y., Oberle, C., & Kassaify, Z. (2003). Eggshell matrix proteins as defence mechanism of avian eggs. *Journal of Agriculture and Food Chemistry, 51*(1), 249–253. https://doi.org/10.1021/jf020597x

Mittal, A., Teotia, M., Soni, R. K., & Mittal, J. (2016). Applications of egg shell and egg shell membrane as adsorbents: A review. *Journal of Molecular Liquids, 223*, 376–387. https://doi.org/10.1016/j.molliq.2016.08.065

Nakano, T., Ikawa, N. I., & Ozimek, L. (2003). Chemical composition of chicken eggshell and shell membranes. *Poultry Science, 82*, 510–514.

Oda, J., Sakai, S., & Kemmochi, S. (1998). Mechanical evaluations of structural and material composition of eggshell. *JSME International Journal Series A Solid Mechanics and Material Engineering, 41*(1), 121–126. https://doi.org/10.1299/jsmea.41.121

Park, S., & Kim, J. (2017). Eggshell membrane: Review and impact on engineering. *2017 ASABE Annual International Meeting*. https://doi.org/10.13031/aim.201700737

Park, S., Choi, K. S., Lee, D., Kim, D., Lim, K. T., Lee, K. H., Seonwoo, H., & Kim, J. (2016). Eggshell membrane: Review and impact on engineering. *Biosystems Engineering, 151*, 446–463. https://doi.org/10.1016/j.biosystemseng.2016.10.014

Pereira, J. F. B., & Coutinho, J. A. P. (2020). *Chapter 5: Aqueous two-phase system. in liquid-phase extraction*, 157–182. Elsevier. https://doi.org/10.1016/B978-0-12-816911-7.00005-0

Poland, A. L., & Sheldon, B. W. (2001). Altering the thermal resistance of foodborne bacterial pathogens with an eggshell membrane waste by-product. *Journal of Food Protection, 64*(4), 486–492. https://doi.org/10.4315/0362-028X-64.4.486

Preito, J. M. (2012). Preparation of DPPH Radical, and antioxidant scavenging assay. DPPH Microplate protocol, 7–9.

Raja, S., Murty, V. R., Thivaharan, V., Rajasekar, V., & Ramesh, V. (2011). Aqueous two-phase systems for the recovery of biomolecules: A review. *Science and Technology, 1*(1), 7–16. https://doi.org/10.5923/j.scit.20110101.02

Rodriguez-Navarro, A. B., Yebra, A., Nys, Y., Jimenez-Lopez, C., & Garcia-Ruiz, J. M. (2007). Analysis of avian eggshell microstructure using X-ray area detectors. *European Journal of Mineralogy, 19*(3), 391–398. https://doi.org/10.1127/0935-1221/2007/0019-1724

Rose-Martel, M., Smiley, S., & Hincke, M. T. (2015). Novel identification of matrix proteins involved in calcitic biomineralization. *Journal of Proteomics, 116*, 81–96. https://doi.org/10.1016/j.jprot.2015.01.002

Ruff, K. J., DeVore, D. P., Leu, M. D., & Robinson, M. A. (2009). Eggshell membrane: A possible new natural therapeutic for joint and connective tissue disorders. Results from two open-label human clinical studies. *Clinical Interventions in Aging, 4*, 235–240. https://doi.org/10.2147/cia.s5797

Santana, A., Melo, A., Tavares, T., & Ferreira, I. M. P. L. V. O. (2016). Biological activities of peptide concentrates obtained from hydrolysed eggshell membrane byproduct by optimisation with response surface methodology. *Food and Function, 7*(11), 4597–4604. https://doi.org/10.1039/c6fo00954a

Shi, Y., Kovacs-Nolan, J., Jiang, B., Tsao, R., & Mine, Y. (2014). Antioxidant activity of enzymatic hydrolysates from eggshell membrane proteins and its protective capacity in human intestinal epithelial Caco-2 cells. *Journal of Functional Foods, 10*, 35–45. https://doi.org/10.1016/j.jff.2014.05.004

Shi, Y., Zhou, K., Li, D., Guyonnet, V., Hincke, M. T., & Mine, Y. (2021). Avian eggshell membrane as a novel biomaterial: A review. *Foods, 10*(9), 2178. https://doi.org/10.3390/foods10092178

Sim, W. J., Ahn, J., Lim, W., Son, D. J., Lee, E., & Lim, T. G. (2023). Anti-skin aging activity of eggshell membrane administration and its underlying mechanism. *Molecular and Cellular Toxicology, 19*, 165–176. https://doi.org/10.1007/s13273-022-00291-5

Statista. (2022a). CAP: Malaysians deserve safe eggs too. Retrieved from https://consumer.org.my/cap-malaysians-deserve-safe-eggs-too/

Statista. (2022b). Egg Production Worldwide since 1990. Retrieved from https://www.statista.com/statistics/263972/egg-production-worldwide-since-1990/

Statista. (2022c). Malaysia-Quantity of Chicken Eggs Produced. Retrieved from https://www.statista.com/statistics/1318592/malaysia-quantity-of-chicken-eggs-produced/

Statista. (2022d). Top 10 Countries Worldwide in Egg Production. Retrieved from https://www.statista.com/statistics/263971/top-10-countries-worldwide-in-egg-production/

Tan, J. S., Abbasiliasi, Y. K., Lin, M. S., Mohamed, M. R., Kapri, M. R., Kadkhodaei, S., Tam, Y. J., Rahman, R. N. Z., & Ariff, A. B. (2014). Primary recovery of thermostable lipase 42 derived from recombinant *Escherichia coli BL21* in aqueous two-phase flotation. *Separation and Purification Technology, 133*, 328–334. https://doi.org/10.1016/j.seppur.2014.06.048

Tapal, A., & Tiku, P. K. (2019). Nutritional and nutraceutical improvement by enzymatic modification of food proteins. In *Enzymes in Food Biotechnology*, 471–481. Academic Press. https://doi.org/10.1016/B978-0-12-813280-7.00027-X

Toma, M., Vinatoru, M., Paniwnyk, L., & Mason, T. J. (2001). Investigation of the effects of ultrasound on vegetal tissues during solvent extraction. *Ultrasonic Sonochemistry, 8*(2), 137–142. https://doi.org/10.1016/S1350-4177(00)00033-X

Torres, F. G., Troncoso, O. P., & Montes, M. R. (2013). The effect of temperature on the mechanical properties of a protein-based biopolymer network. *Journal of Thermal Analysis and Calorimetry, 111*(3), 1921–1925. https://doi.org/10.1007/s10973-012-2915-0

Torres, F. G., Troncoso, O. P., Piaggio, F., & Hijar, A. (2010). Structure-property relationships of a biopolymer network: The eggshell membrane. *Acta Biomaterialia, 6*(9), 3687–3693. https://doi.org/10.1016/j.actbio.2010.03.014

Tsai, W. T., Yang, J. M., Lai, C. W., Cheng, Y. H., Lin, C. C., & Yeh, C. W. (2006). Characterization and adsorption properties of eggshells and eggshell membrane. *Bioresource Technology, 97*(3), 488–493. https://doi.org/10.1016/j.biortech.2005.02.050

United Nations. (2022). SDG 12: Ensure sustainable consumption and production patterns. Retrieved from https://sdgs.un.org/goals/goal12

Xiao, F., Xu, T., Lu, B., & Liu, R. (2020). Guidelines for antioxidant assays for food components. *Food Frontiers, 1*, 60–69. https://doi.org/10.1002/fft2.10

Yi, F., Yu, J., Guo, Z. X., Zhang, L. X., & Li, Q. (2003). Natural bioactive material: A preparation of soluble eggshell membrane protein. *Macromolecular Bioscience, 3*(5), 234–237. https://doi.org/10.1002/mabi.200390030

Yoo, J., Park, K., Yoo, Y., Kim, J., Yang, H., & Shin, Y. (2014). Effects of egg shell membrane hydrolysates on anti-inflammatory, anti-wrinkle, anti-microbial activity and moisture-protection. *Korean Journal for Food Science of Animal Resources, 34*(1), 26–32. https://doi.org/10.5851/kosfa.2014.34.1.26

Yusof, F. (2011). Protein extraction and purification. In *Experimental Methods in Modern Biotechnology Engineering,* 11–24, IIUM Press. ISBN No: 978–967–0225–86–9.

Yusof, F. (2015). Purification of recombinant protein for industrial use. In *Recombinant Enzymes: From basic sciences to Commercialization (Azura Amid, Ed.),* 61–80, ISBN: 978–3–319–12396–7, ISBN:978–3–319–12397–4 (eBook), Springer International Publishing Switzerland. https://doi.org/10.1007/978-3-319-12397-4

Zhao, Q. C., Zhao, J. Y., Ahn, D. U., Jin, Y. G., & Huang, X. (2019). Separation and identification of highly efficient antioxidant peptides from eggshell membrane. *Antioxidants, 8*(10), 495. https://doi.org/10.3390/antiox8100495

Zhou, J., Wang, S., Nie, F., Feng, L., Zhu, G., & Jiang, L. (2011). Elaborate architecture of the hierarchical hen's eggshell. *Nano Research, 4*(2), 171–179. https://doi.org/10.1007/s12274-010-0067-8

Zhu, L., Ma, M., Ahn, D. U., Guyonnet, V., Wang, L., Zheng, Y., He, Q., Xiong, H., & Huang, X. (2022). Hatched eggshell membrane can be a novel source of antioxidant hydrolysates to protect against H_2O_2-induced oxidative stress in human chondrocytes. *Antioxidants, 11*(12). https://doi.org/10.3390/antiox11122428

Pharmaceuticals

Synthetic Biology as a Building Block for Circular Bioeconomy: A Case Study of Chitin

Fazia Adyani Ahmad Fuad, Suriyea Tanbin, Nur Aqilah Husna Azizi, and Tengku Rogayah Tengku Abdul Rashid

Abstract Synthetic biology has revolutionized many aspects of life, by engineering organisms to have new abilities to be applied for practical uses. Among the fields that benefit most from the advent of synthetic biology is the pharmaceutical field, notably drug discovery. The drug discovery pipeline is known to take over a decade to reach the market, however, with the aid of synthetic biology, the process may well be sped up to completion. This chapter discusses an important aspect of synthetic biology that serves as the basis of circular bioeconomy from a pharmaceutical point of view, which is in silico approach to study protein-drug interactions in novel drug development. This revolves around the analyses of chitin, which is a naturally occurring compound and a biodegradable polymer that has potential as pharmaceutical products and potent drugs that can be obtained at low cost from food and forestry waste, respectively.

Keywords Chitin · In silico analysis · Synthetic biology · Circular bioeconomy

1 Introduction

Synthetic biology is an emerging field that has now become the essence of bioeconomy. This involves creating novel bimolecular components, networks, pathways and subsequently using these constructions to rewire and reprogram organisms. Biotechnology, genetic engineering, molecular biology, molecular engineering, systems biology, chemical and biological engineering, and evolutionary biology are just a few of the disciplines that are represented in the multidisciplinary field of synthetic biology. Beginning with the turn of the twentieth century, the word 'synthetic biology' appeared for the first time in isolation. For instance, Stephane

F. A. A. Fuad (✉) · S. Tanbin · N. A. H. Azizi
Department of Chemical Engineering and Sustainability, Kulliyyah of Engineering, International Islamic University of Malaysia (IIUM), Jalan Gombak, 53100 Kuala Lumpur, Malaysia
e-mail: fazia_adyani@iium.edu.my

T. R. T. A. Rashid
Electron Microscopy Unit, Special Resource Centre, Institute for Medical Research, National Institutes of Health, Ministry of Health Malaysia, 40170 Setia Alam, Selangor, Malaysia

F. Adyani Ahmad Fuad et al. (eds.), *Circular Bioeconomy*,
https://doi.org/10.1007/978-981-97-7010-6_2

Leduc, a French biologist and medical doctor, wrote a book named 'La Biologie Synthétique' in 1912. Leduc argued that in addition to the methods of analysis, evidence gathering, and classification, science also employs a synthetic technique that seeks to replicate observed events in a way that is rule-governed and repeatable. Similar to how synthetic chemistry changed the world a century ago, synthetic biology appears to be on track to do the same for the world nowadays.

Although the term 'synthetic biology' has no universally accepted meaning, its main characteristics include the 'de novo' synthesis of genetic material and an engineering-based approach to the development of parts, organisms, and products. Building contemporary biotechnology procedures and methods, such as high throughput DNA technologies and bioinformatics is also part of synthetic biology. The driving forces of biology have shifted from discovery and observation to hypothesis and synthesis, as a result of synthetic biology (Kitney and Freemont 2012; Lim et al. 2012). The majority of synthetic biology research focuses on industrial and commercial uses, while it may uncover information about the 'origin of life' and further our knowledge of the fundamental roles played by genomes. In order to separate this developing field of study from conventional genetic engineering, the phrase 'synthetic biology' was coined in the early 2000s (Joshi and Hanson 2024). Medicine, specialty or fine chemicals, fuels and fuel additives, plastics, polymers, and rubbers, as well as plant feedstocks for microbe consumption, were ranked as the top five application areas for designers and manufacturers performing synthetic biology research (Woodrow Wilson International Center for Scholars 2013). The summary of the application of synthetic biology and their examples is shown in Fig. 1.

One of the important aspects of synthetic biology is its significance in the discovery of therapeutic drug targets. The drug discovery and development pipeline involves several steps including (1) identifying candidates (i.e. the target and lead compound), (2) preclinical development, (3) clinical trials, (4) gaining regulatory approval, and (5) marketing (Biala et al. 2023). Prior to developing a suitable drug, basic research is

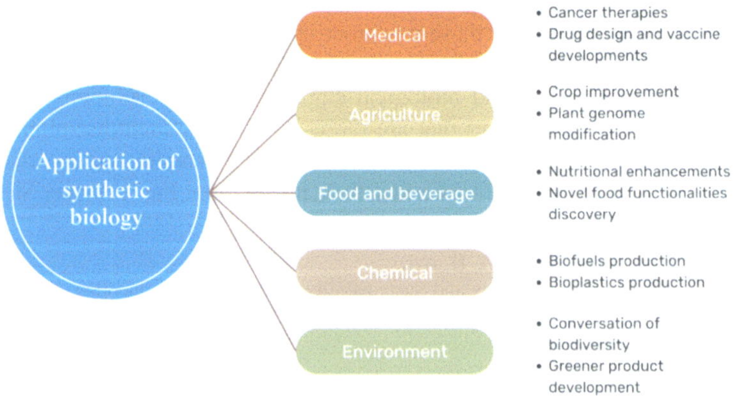

Fig. 1 The application of synthetic biology

needed to identify and select the target, which is a macromolecule such as proteins, an abnormal signaling pathway, or a biological process that appears associated with the desired medical condition (Singh et al. 2023). An ideal target must be effective, safe, satisfy clinical and commercial objectives, and be 'druggable' (Deore et al. 2019). According to Hughes et al. (2011), upon binding to the alleged lead compound, a 'druggable' target should elicit a biological response that can undergo assays both in vitro and in vivo. Few methods can be applied during this step, however, it must be based on the principles of biochemistry, molecular biology, genetics, and other related disciplines.

Following the target's identification, scientists must also identify a therapeutic substance or lead compound that alters the disrupted target's function to recuperate health or relieve symptoms (Chen et al. 2018; Singh et al. 2023). This substance will undergo a few tests and experiments to obtain its chemical properties, toxicity, effectiveness, mechanism of action, and the perfect dosage to be administered to humans (Chen et al. 2018; Food and Drug Administration 2018b). For it to be termed 'druggable', a compound must have the ability to bind to a specific target and exhibit a favorable pharmacokinetic profile, including absorption, distribution, metabolism, and excretion rates (Deore et al. 2019). During lead discovery, intense research is carried out in search of a drug-like small molecule or biological treatment, known as a development candidate, that will advance into preclinical evaluation, clinical trials, and, if successful, commercialization (Hughes et al. 2011). Traditionally, finding lead compounds will involve numerous experiments and 'trial and error' in the laboratory. However, currently, researchers opt for computational or in silico analysis since this tool enables quick identification of protein–ligand binding sites by screening multiple virtual databases.

In drug discovery, the knowledge of protein–ligand interactions is crucial in facilitating the discovery of novel therapeutic drugs. Traditional studies of protein–ligand interaction involve experiments in the laboratory. Nevertheless, this approach is time-consuming, costly, and sometimes inaccurate (Zhang et al. 2021). Hence, researchers nowadays opt for a more advanced technique of drug discovery which is in silico or computational method, since it has considerably reduced the time and costs. In silico methods for investigating the protein–ligand interactions and determining protein functionality are favorable since sequence databases are rapidly expanding, making in vivo functional analysis insufficient, laborious, and time-consuming (Roche and McGuffin 2016). According to Yang et al. (2022), Computer-Aided Drug Discovery (CADD) has been demonstrated to be a potential method for advancing the research and development of new drugs. There are a few other essential ways for studying protein–ligand interactions when it comes to in silico investigation, which includes virtual screening (VS) and machine learning (ML).

Virtual screening (VS) analysis has been widely employed in the research of novel drugs. VS is a type of computational method that helps in predicting and selecting a strong binding compound for a protein of interest from huge ligand databases (Shin and Kihara 2018). The most employed methods under VS are the ligand-based and structure-based methods. The ligand-based method, also known as Ligand-Based Drug Discovery (LBDD) technique helps in identifying novel ligands that can bind to the target protein's structure (Ajjarapu et al. 2021). LBDD evaluates present research using methodologies including quantitative structure–activity relationship (QSAR) models, chemical similarity, pharmacophore, and 3D shape matching to predict the properties of a novel compound (Yang et al. 2022).

Structure-based method, also known as Structure-Based Drug Discovery (SBDD), is frequently employed to assist researchers in predicting the location of small molecules inside a three-dimensional (3D) model of the protein structure and estimating ligand affinity to target proteins with high accuracy and efficiency (Wang et al. 2018). Unlike LBDD, SBDD only relies on the 3D structure of the molecule compound (Jung et al. 2023). One of the most popular techniques in SBDD is molecular docking (Batool et al. 2019). The molecular docking method is a virtual simulation method that works to predict the conformation and binding affinities of a compound with ligand through multiple scoring functions (Agu et al. 2023). Both of these techniques, i.e. SBDD and LBDD, can also be combined for use in virtual screening drug discovery.

However, as years passed, a new method of VS called the network-based method has been developed to accommodate the demand for new therapies by predicting the protein–ligand interactions via biological networks and graph theory (Wu et al. 2022; Zhao et al. 2022). In contrast to SBDD, this approach does not involve a 3D structure of the target or negative sample, which renders it simple and fast. Moreover, this approach is not limited by the 3D structure of targets, allowing it to incorporate a much larger target space (Wu et al. 2018). Furthermore, this method is also capable of predicting potential interactions between target and lead compounds by employing straightforward physical operations such as resource diffusion, collaborative filtering, and random walk-on networks (Wu et al. 2018).

Once the lead compound has been identified, the preclinical formulation and development stage will be conducted prior to clinical trials to investigate the preparation method, dosage, toxicity, safety, efficacy, and others (Chen et al. 2018; Deore et al. 2019). Pre-clinical studies can be undertaken in two types: general pharmacology and toxicology (Deore et al. 2019). Pharmacology examines the pharmacokinetic and pharmacodynamic properties of drugs. Meanwhile, pharmacokinetic studies of a drug are crucial in determining the drug's effectiveness and safety dosage intake, metabolism, and disposal (Deore et al. 2019).

Before commencing clinical trials on the drug, an Investigational New Drug (IND) will be filed to regulatory bodies such as the Food and Drug Administration (FDA) and the National Pharmaceutical Regulatory Agency (NPRA) (Singh et al. 2023). This IND acts as an application by a clinical trial sponsor to acquire authorization from regulatory bodies to administer an experimental drug or any biological substance to human beings (Food and Drug Administration 2022). Clinical trials on humans will begin once the IND has been approved, which are divided into four phases. During the first phase, the drug's safety and tolerability are investigated in a small number of healthy people or volunteers around 20 to 100 people (Food and Drug Administration 2018a; Singh et al. 2023). Phase 2 will feature a maximum of several hundred individuals with the disease to examine its efficacy and adverse effects. This phase will take up to several months to two years with approximately 33% of drugs going to the subsequent phase.

In the third phase, more than 300 patients will be recruited to evaluate the efficacy and safety of the drug and compare the test drug to the placebo (Kandi and Vadakedath 2023). After completing the phase three study, a New Drug Application (NDA) will be submitted to regulatory bodies for demonstration of drug safety and efficacy (Singh et al. 2023). Lastly, under numerous instances, regulatory authorities will request further follow-up investigations, generally called phase four or post-marketing monitoring of indefinite duration to discover uncommon detrimental impacts or long-term unfavorable effects over a much larger patient group. After completing all four phases of clinical trials, the new drug will undergo regulatory approval and be developed for market release.

Overall, synthetic biology has the potential to revolutionize the drug discovery and development process as it enables researchers to synthesize new and more effective drugs, taking the instance of the in silico studies for the investigation of protein–ligand interactions. Since technological advancements have made it possible to engineer biological systems, it also allows scientists to develop new bimolecular components, networks, and pathways, which can be engineered to reprogram organisms to synthesize therapeutic agents. By synthesizing new and more accurate drugs, researchers can also find drugs that are more specific and result in fewer side effects. Besides, synthetic biology can also be used to develop diagnostic tools that help scientists detect diseases faster and more accurately. Therefore, synthetic biology is likely to revolutionize drug development and better patient outcomes. Figure 2 below shows the summary of the stages of novel drug discovery and development.

The case study in this chapter will employ molecular docking which falls under the SBDD technique in estimating the binding affinity between the target protein with a small molecule ligand. SBDD is often favored when high-resolution structural data for the target protein are readily available. Furthermore, in comparison with free energy calculations, SBDD can be easily interpreted into experimental data that allows researchers to further the experimental work. The information obtained from this method can provide insights into the binding affinity, stability, and specificity of the protein–ligand complex. This can aid in the application of synthetic biology in the design of new drugs by predicting the binding mode and identifying potential drug candidates that can bind to the target protein with high affinity and specificity. Overall, in silico methods have significantly contributed to the drug discovery process in recent years.

This chapter will highlight the application of synthetic biology in the case study of chitin as a ligand of interest. In the case study, the targeted protein for the discovery of dengue fever drug is the hexokinase isoform II (HKII). HKII is an enzyme that is involved in the first step of the glycolytic pathway and has been shown to exhibit a potential target for cancer treatment (Garcia et al. 2019). Studying and understanding the role of HKII as well as utilizing this enzyme to create targeted drugs that can suppress its activity rely highly on synthetic biology. In addition to that, the intersection of synthetic biology and circular bioeconomy has also become a hot topic these days along with the implementation of Sustainable Development Goals (SDG).

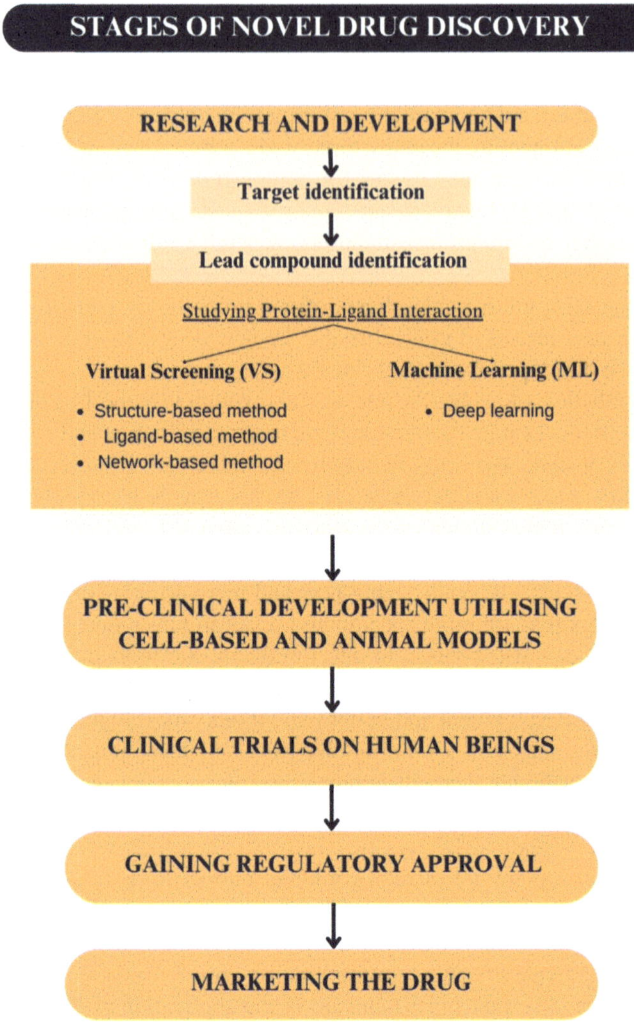

Fig. 2 The stages of novel drug discovery and development (Biala et al. 2023)

Hence, this chapter will commence a voyage of the convergence of synthetic biology and circular bioeconomy specifically on how synthetic biology can contribute to the enhancement of the circular bioeconomy, by focusing on SDG3: Good Health and Well-being.

2 The Bioeconomy is Being Driven by Synthetic Biology

The bioeconomy essentially revolves around the production of standard goods and services utilizing renewable feedstocks. The Organization for Economic Cooperation (OECD) 2009 report; The Bioeconomy to 2030: Designing a Policy Agenda established some restrictions on the notion of the bioeconomy, but it has since grown significantly. It now includes a wide range of industries and activities, such as waste management, chemicals, food, agriculture, dairy, and forestry. Biotechnology is only one aspect of the bioeconomy. In line with the idea of a circular economy, it is now viewed as a new method of production that will gradually replace fossil-based manufacturing (Philp and Winickoff 2018). The manufacturing of high-value compounds from sustainable sources with less energy and waste is now possible, thanks to biotechnology, which makes use of enzymatic processes. An early illustration of this would be the enzymes in laundry detergent, which enable effective cleaning at lower temperatures (i.e. with reduced energy use). Enzymatic catalysis is now being utilized to produce entire species, ranging from bacteria to yeast to plants. Complex and sought-after molecules can be created simply using glucose as a starting material, from the creation of insulin in genetically altered bacteria to the synthesis of artemisinic acid—a malaria medication precursor in designed yeast. White biotechnology has benefited greatly from synthetic biology, which applies engineering ideas to biology and allows for the rational design or enhancement of production routes. White biotechnology is ideally positioned to serve as the cornerstone for a circular bioeconomy, with synthetic biology being hailed as one of the ten most significant emerging technologies by the World Economic Forum (2016) and biotechnology being named one of the six most important technologies for our sustainable future by the EU. The synthetic biology biodesign cycle and impacts on circular bioeconomy are shown in Fig. 3.

Synthetic biology makes use of existing genetic modification technologies with the specific goal of accelerating and facilitating the process, which includes making it more predictable (European Community 2014). DNA-based circuits, synthetic metabolic pathway engineering, synthetic genomics, protocell assembly, and xenobiology are some of the fields of study that fall under the umbrella term 'synthetic biology'.

3 Synthetic Biology Speeds up Drug Target Discovery

Synthetic biology is a rapidly developing area that aims to create artificial synthetic circuits that precisely regulate cellular networks (Xie 2020). This technology has significant potential for the treatment of disease and the identification of new therapeutic targets. Different synthetic circuits are made to implement multilayer rules

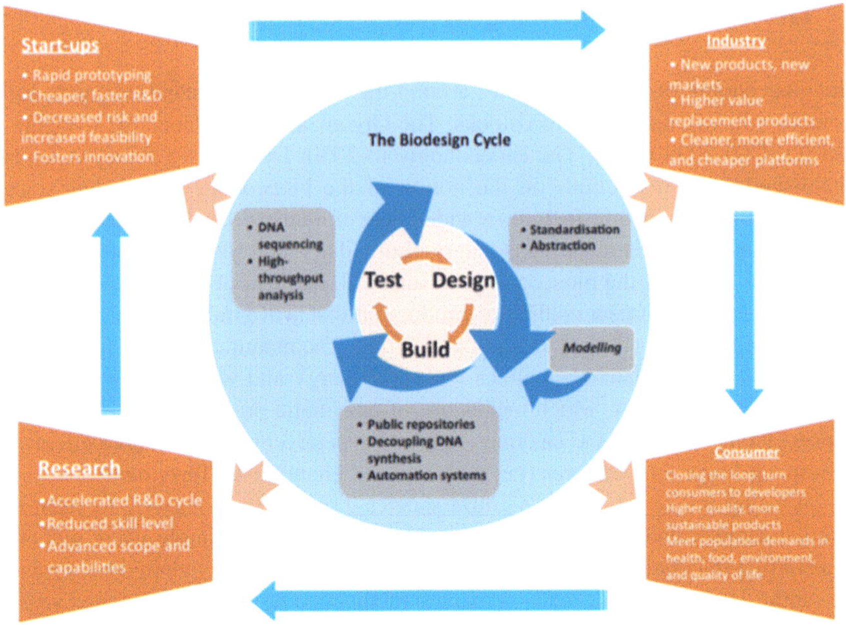

Fig. 3 The synthetic biology biodesign cycle and its impact on circular bioeconomy

depending on the combination mode of distinct logic gates. Drugs frequently function as inputs to drive circuit operation in specific synthetic circuits. Building drug-responsive gene circuits is becoming possible for treating numerous disease models experimentally, including cancer, bacterial and viral infection, immune disease, and metabolic disease. For effective drug target screening, synthetic biology and the Clustered Regularly Interspaced Short Palindromic Repeats (CRISPR) system work well together. Surprisingly, more and more expertly crafted circuits are being created to identify brand-new drug targets and accurately control drug therapy for illnesses. In the recent years, the laboratory of Christopher Voigt has started collaborating with experts from the pharmaceutical industry to assess the usefulness of synthetic biology for drug discovery (Drug Discovery World 2016). The synthetic biology market is estimated to be worth $5 billion globally by a number of market research studies, with up to 35% compound annual growth predicted over the coming years. One of the key factors influencing the growing demand is the interest from pharmaceutical and biotechnology businesses (Drug Discovery World 2016).

The genomic information that has been gathered by the community may be analyzed using bioinformatics tools to identify the pathways that produce a variety of potentially intriguing substances, such as tiny molecules produced in the gut that are interacting with the host in some way. Prior to the advent of synthetic biology, it was impossible to exploit metagenomics sequence data for drug development via accessing pathways. Scientists synthesized the genes or pathways and transported them into a new host for study due to DNA synthesis, a process that underpins much of synthetic biology. Even if the organism has long since vanished, native DNA sequences still can be accessed and examined, recodifying the DNA to make it compatible with a different host. Using synthetic biology, one can build the pathway for optimum function and regulate the timing and levels of expression. An important approach to leverage the investment in sequencing databases and identify new biological activities for application in medicine is to be able to delve into genetic databases to find DNA responsible for intriguing new molecules. The tools underlying synthetic biology have just recently improved to enable this field to be used in areas like medication discovery and development. Synthetic biology is now prepared for application on an industrial scale, thanks to advancements in gene synthesis, bioinformatics, and the number of biological components available. It is important to keep in mind that synthetic biology is still in a relatively early stage compared to disciplines like organic chemistry, where researchers can reliably discover a synthesis path to almost any chemical. The ability to create any desired chemical structure by fusing enzymes from other animals is a major challenge for synthetic biology, and still a long way from attaining it. A biological pathway is particularly suitable for many complex compounds, such as those that include several functional groups or stereocenters. Additionally, under suitable environmental circumstances, cells can produce the compound in high yields. The process of synthetic biology doesn't really start until a proper chemical is found, whether it is in one's own guts or in a genetic database. It takes a lot of work to recode the DNA, choose the best codons for optimal gene expression, and streamline the sequence to avoid introducing any undesired processes along the way in order to successfully engineer the genes or pathways into a host organism. By using metabolic engineering, a rigorous, large-scale manufacturing of the molecule is made possible. Drug research is being transformed into an information science thanks to synthetic biology. Scientists will search databases for interesting genes or proteins instead of prospecting for novel substances in tropical forests. They will then test each gene or protein many times to find the best candidate. The transition from serendipity to rational design in drug discovery will enable us to develop safer, more potent treatments with fewer late-stage clinical trial failures in the future.

4 Chitin as a Potential Drug for Disease Treatment and Its Applications—A Case Study

After cellulose, chitin is the most prevalent natural polymer in the world. It is mostly made from the leftover meal of prawns and crabs. Sponge, coral, mollusc, worm, and arthropod species, as well as unicellular (diatoms, protists, fungi) and multicellular (coral, mollusc, worm, and arthropod) animals, have all been experimentally proven to contain chitin, which is a long-chain polymer of N-acetylglucosamine (Tsurkan et al. 2021). Chitin (Fig. 4) is a glycosidic-bonded linear polymer that contains -(1,4)-N-acetyl-d-glucosamines. Chitin comes in three different isoforms:

-α, -β, and -γ chitin (Moussian et al. 2019). The arrangement of the side chains of the backbone determines the various chitin types. The exoskeleton of arthropods and other organisms that regularly occur in nature include β-chitin, which has a parallel chain arrangement as its structure. The use of chitin in tissue engineering is common (Jayakumar et al. 2011; Kumar et al. 2011). Among biopolymers that are biodegradable and made from renewable raw materials, chitin and chitosan are widely used in many sectors such as biomedical, biotechnology, water treatment, food, agriculture, veterinary, and cosmetics. The majority of chitin, which is α-chitin, may be found in squid pens and has an antiparallel chain arrangement in its structure. Additionally, β-chitin has been used in biomaterials for things like methylene blue preservation, wound healing, and biological applications including enhancing salty perception (Song et al. 2019; Qin et al. 2019). The γ-chitin has been extracted from the cocoon of the moth (*Orgyia dubia*), and it has been discovered that its structure is similar to α-chitin rather than β-chitin, but that it has a different surface morphology with microfibers instead of nanofibers (Song et al. 2019). Chitin can be produced at high temperatures and has a high degree of thermostability. Additionally, because chitin is very tolerable to high chemical concentrations, metals like copper can be electrochemically deposited onto chitin at room temperature (Pertrenko et al. 2017). According to Petrenko et al. (2017) and Khrunyk et al. (2020), chitin has been employed as an additive for medicinal products, nutraceutical food, and 3D scaffolds for biomedical and technical applications.

It is widely known that when chitin enters the human body, chitinase, an enzyme that regulates immunological response, and chitinase-like proteins YKL-40 are produced. Tang et al. (2010) reported that YKL-40 was linked to the worsens of asthmatic patients' symptoms. According to Liu et al. (2019), there was a positive correlation between YKL-40 level and neutrophils, sputum IL-1, and plasma IL-6. In a mouse model, chitin from fungi also caused eosinophilic infiltration. In addition, human bronchial epithelial cells that were exposed to chitin were stimulated to generate proinflammatory cytokines such IL-25 and IL-33 (Khosravi and Erle. 2016). Chitin has a history of being used as an immune response adjuvant. A TLR-2 dependent pathway was shown to be used by house dust mite chitin to increase airway hypersensitivity in ovalbumin-induced airway inflammation (Choi et al. 2016). Furthermore, it is well known that the main element of fungi's cell walls is chitin.

According to Boot et al. (1998), chitin exposure caused macrophage activation, which increased the production of the chitin-degrading enzyme chitotriosidase. Inhibition of chitotriosidase and acidic mammalian chitinase is viewed as a therapeutic target for respiratory disorders because these enzymes were also engaged in lung diseases like tuberculosis and chronic obstructive pulmonary diseases (Mazur et al. 2018). Chitin has been demonstrated to have therapeutic benefits when coupled with other substances. For instance, oral chitin and glucan intake of 4.5 g/day in human participants decreased oxidized low-density lipoprotein (Bays et al. 2013). Chitin has also been used as a biomaterial to extend bioavailability. For example, to extend long-term sustained salmon calcitonin release, an injectable thermo-sensitive hydrogel based on hydroxypropyl chitin was added (Yu et al. 2020).

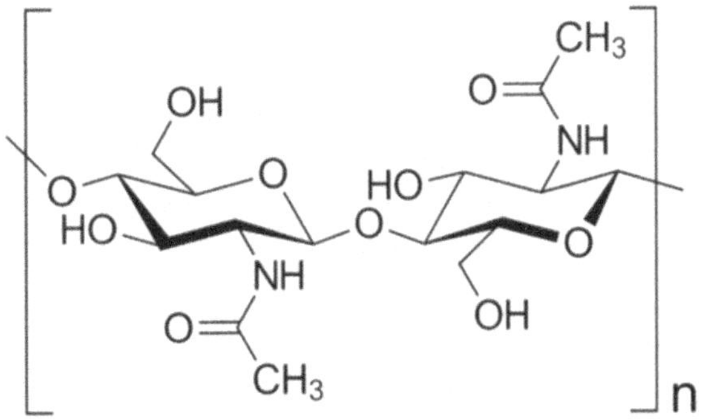

Fig. 4 The chemical structure of chitin

Chitin is non-toxic and biodegradable, making it a desirable substance for use as a biomaterial in medicine. In applications such as tissue engineering, medication delivery, and hemostatic dressings, it has been continuously studied. The interest in chitin has extended to nanostructures; research on nanofibers and nanoparticles for biological purposes is ongoing, with the latter receiving special attention for its potential utility in drug administration. Chitin and its derivatives are known to efficiently suppress the growth of both gram-positive and gram-negative bacteria, a property attributed to disruptions to bacterial membranes by the amine group present in the molecule. This antibacterial property of chitin and its derivatives is part of the reason for this interest. Although it inhibits the growth of gram-positive bacteria, chitin may encourage the growth of gram-negative bacteria. The usage of chitosan in biomaterials has increased due to its broad spectrum of antibacterial characteristics and chitin's insoluble nature. For use in treating wounds, chitin acts as hemostats and has generated interest, where chitosan gauze is readily accessible in the marketplace. The goal of recent research in this field is to develop composite materials that either enhance hemostatic capabilities or take advantage of them to speed up tissue regeneration in injured areas. Researchers have invented novel materials that may enhance wound treatment which is composite gauzes from chitosan and alginate and then compared them to gauzes with a catechol surface modification. According to He et al. (2022), chitosan-catechol gauze has shown outstanding hemostatic properties in rat femoral artery and liver laceration injury models. In fact, the blood loss from this modified chitosan gauze was discovered to be less than 17% of that of pristine chitosan gauze (He et al. 2022). Additionally, the hemostatic function of chitosan microspheres was further enhanced by coating them with calcium alginate or zinc alginate to create core-sheath microspheres, since Ca^{2+} and Zn^{2+} can increase the formation of thrombin (He et al. 2022).

Many of the same qualities that make chitin and its derivatives desirable for tissue engineering also make them suitable for the usage in medication administration. Chitin is now easier to use for drug administration, especially in the form of hydrogels, due to advancements in ionic liquids and deep eutectic solvents (DESs) (Li et al. 2022). Owing to its solubility and more extensive anti-microbial activity, chitosan is generally more appealing as mentioned above. For a variety of delivery techniques, including oral, nasal, intravenous, and transdermal channels, among others, chitin-based drug carriers have been created (Desai et al. 2023). Chitosan may take on a broad variety of morphologies, and it is stable when conjugated with other common components used in drug carriers, which contributes to its flexibility. Since it has the capacity to conjugate, chitosan is appealing for drug delivery since it can be functionalized to enhance the targeted release of medications (Qin and Li 2020). To prevent toxicity from spreading to other parts of the body, current drug delivery research focuses on creating non-toxic, biodegradable carriers that release the medications contained in a specified area.

Chitin also acts as antioxidants, protecting skin from oxidative damage when incorporated into cosmetics (Kulka and Sionkowska 2023). Due to the positive charge of chitosan, it can act as a humectant to maintain skin moisture or strengthen hair and as a conditioner through interactions with hair keratin (Aranaz et al. 2018). Suspensions of chitosan nanoparticles in guar gum have been indicated to reduce skin sebum levels in seborrhea patients (Theerawattanawit et al. 2021). Chitin nanofibrils electro sprayed onto cellulose films improved compatibility with skin cells and downregulated inflammatory cytokines, indicating potential applications for nanochitin in cosmetics (Azimi et al. 2021). Among the polymers, chitin has shown particular interest because it is biodegradable, non-toxic, skin- and environmentally friendly ingredients can be obtained at low cost from food and forestry waste, respectively.

5 In Silico Approach for Studying Protein–Ligand Interactions in Novel Drug Development

Investigating protein–ligand interactions is an important initial step in the development of novel drugs. Drug development consists of three main phases: (1) drug discovery, (2) preclinical development utilizing cell-based and animal models, (3) clinical trials on human beings, and (4) gaining regulatory approval to market the drugs (Biala et al. 2023). Protein–ligand interaction studies are done as the preliminary steps of drug development which is in the drug discovery phase. Previously, investigations of protein–ligand interactions were conducted using time-consuming and difficult in vitro experimental methods such as high throughput screening (HTS). Nevertheless, with advancements in computers and technology, drug discovery presently depends heavily on computational approaches or in silico screening, which is a more effective approach in many instances.

In living organisms, proteins are unable to function independently; they have to be linked to other molecules to establish a particular interaction to fulfill their purpose, and these molecules are called ligands (Dhakal et al. 2022). Protein–ligand interaction is crucial in the development of novel drugs to predict the binding affinity of the protein of interest with a ligand, in terms of finding the suitable lead compound. This binding affinity will then show the strength of the interactions between the lead compound and the target. Abnormal interactions of protein–ligand may result in numerous medical conditions, hence it is crucial to have full knowledge of the processes behind these interactions in order to facilitate the development of novel therapies (Wang et al. 2023).

In predicting the protein–ligand interaction, the in silico method can be divided into two: (1) virtual screening (VS) and (2) machine learning (ML) and deep learning (DL) (Kimber et al. 2021). VS is a method utilized to rank compounds from an enormous compound database that have an excellent likelihood of binding to a target of interest (Kimber et al. 2021). The most common methods of VS are ligand-based and structure-based methods. However, according to a review by Zhao et al. (2022), to date, three methods of VS can be applied including ligand-base, structure-based, and network-based methods.

In the ligand-based method (LBDD), a single or a collection of compounds that are deemed to bind to a particular target or considered active in a functional assay will typically serve as the template for finding compounds that are similar in huge virtual databases (Jung et al. 2023). To put it simply, the ligand-based technique enables quick screening of massive compound databases to locate similar structures based on their 2D or 3D structures (Jung et al. 2023). Since this method only utilizes sets of previously identified ligands for the target of interest, it limits its potential to discover novel chemotypes (Karki et al. 2021).

Compared to the ligand-based method, the structure-based approach solely employs the 3D structure of a specific target to screen libraries, increasing its usefulness in discovering new therapies (Jung et al. 2023). Structure-based virtual screening, also known as target-based virtual screening, can anticipate the optimal interaction of ligands with a molecular target to generate a complex. The most utilized approach is molecular docking, which predicts the binding poses of query ligands and values their binding affinity (Kimber et al. 2021). With this method, ligands will be sorted following their affinity to the receptor site of the molecular target, having the most promising molecules at the top of the list (Maia et al. 2020). The use of this approach is highly popular in the early stages of drug development because of its inexpensive cost, reduced time invested, and a variety of accessible computational tools.

Over the years, new virtual screening techniques, namely the network-based method have been developed to cater to the shift from the linear mode of 'one drug—one target—one disease' to the network mode of 'multiple drugs—multiple targets—multiple diseases' (Wu et al. 2022). This fresh approach implies that a single drug might work on numerous targets in vivo instead of preferentially binding to only one target (Wu et al. 2018). A network-based method is generated from the recommendation algorithms implemented in recommender systems and links the prediction algorithms in complex networks (Wu et al. 2018). As opposed to the structure-based technique, the network-based method is not dependent on the molecular target's 3D structure or any negative samples, which is advantageous (Wu et al. 2022). There are several network-based methods such as network-based inference (NBI), random walk-based methods, and local community-based methods. Nevertheless, these prediction algorithms are shallow-learning methods that are incapable of extracting deep and complicated correlations between compounds and proteins (Zhao et al. 2022).

Other techniques for predicting the protein–ligand interaction are machine learning (ML) and deep learning (DL). DL falls under the umbrella of ML, also known as the subset of ML. In contrast to VS, ML evaluates the relationship between physicochemical properties and protein–ligand interactions using known protein–ligand complex structures to generate statistical models that predict the status of other unknown ligands/proteins (Dhakal et al. 2022). At the beginning of ML, details about the target protein and ligand are collected, followed by data preprocessing by the normalization method (Dhakal et al. 2022). Normalization is commonly employed in ML which involves scaling or transforming the data to ensure that each characteristic contributes equally (Singh and Singh 2020). Normalized data is then supplied into an ML model, several hidden processes will occur, and the expected decision is ultimately generated (Dhakal et al. 2022). Since it requires no explicit hard-coded criteria developed by human professionals to generate predictions and possibly provide high prediction accuracy, there has been a lot of interest in utilizing ML, especially powerful deep learning algorithms, to anticipate protein–ligand interactions (Dhakal et al. 2022). Nevertheless, the effectiveness of existing DL approaches for protein–ligand interactions remains unsatisfactory primarily because of the complexity of the protein–ligand interactions and the insufficient data provided (Zhao et al. 2022). Some DL-based algorithms also fail to make efficient use of massive amounts of unlabeled data (Zhao et al. 2022).

In the case study of this chapter, the in silico method that was employed in the initial step of the study is molecular docking which falls under the structure-based method. Molecular docking helps in the discovery of new compounds of therapeutic significance, predicting protein–ligand interactions at the molecular level, and identifying structure–activity relationships (SAR) (Pinzi and Rastelli 2019). Besides requiring low-cost and effective screening, this method offers a more straightforward and rational approach to drug discovery compared to other in silico screening methods (Meng et al. 2011). The structure-based method is often favored when high-resolution structural data for the target protein are readily available. Furthermore, this method can easily be interpreted into experimental data that allows researchers to further the experimental work. The information obtained from this method can provide insights into the binding affinity, stability, and specificity of the protein–ligand complex. This can aid in the application of synthetic biology in the design of new drugs by predicting the binding mode and identifying potential drug candidates that can bind to the target protein with high affinity and specificity. Overall, in silico methods have significantly contributed to the drug discovery process in recent years.

6 Materials and Methods

Virtual screening is a computer-based technique used in drug design to search novel molecules that are most likely bind to drug target such as protein and enzyme. In this study, virtual screening comprises of ligand-based screening, structure-based screening, and molecular dynamics (MD) simulation.

6.1 Ligand-Based Screening

Ligand-based screening method is used to choose set of active ligands that are able to bind in the active site of target protein. There are plenty of online compound databases, compound libraries, softwares, and web tools where billions of commercial compounds are available for optimization and identification of potential drug-like compounds.

6.1.1 Identification of Chitin

In this study, ligand-based screening was conducted to identify chitin by comparing it to a query molecule 2-Deoxyglucose (2DG) using shape similarity, physiological properties, and toxicity data from extensive compound libraries. This approach involves searching for ligands structurally similar to 2DG that can bind to a receptor, with the screening based on shape similarity and binding affinity. Both 2DG and chitin share common structural features, including a glucose chain with six carbon atoms (Fig. 5). Additionally, the functional feature of 2DG and chitin is they should be able to show an inhibitory effect on human HKII enzyme. The structure data file (SDF) 2DG (ID CHEBI: 17,719), as well as molecular formula ($C_6H_{13}O_9P$) were extracted from PubChem (https://pubchem.ncbi.nlm.nih.gov/) to screen for chitin by using Ultrafast Shape Recognition with CREDO Atom Types (USRCAT) available at http://usr.marseille.inserm.fr/. A total of 100 compounds, which are similar hits of 2DG have been countenanced from the USRCAT program. All predicted molecules were ranked based on a zero (0) to one (1) score, where chitin with a score near to one indicated greater similarity towards 2DG molecule (Table 1). Later, chitin was screened through a structure-based approach and finally analyzed through MD studies (Tanbin et al. 2020).

6.1.2 ADME Analysis

Physiochemical properties, such as absorption, distribution, metabolism, and excretion of chitin and 2DG were predicted from the SwissADME website (http// www.swissadme.ch/index.php) prior to a docking study. The structure data file (SDF)

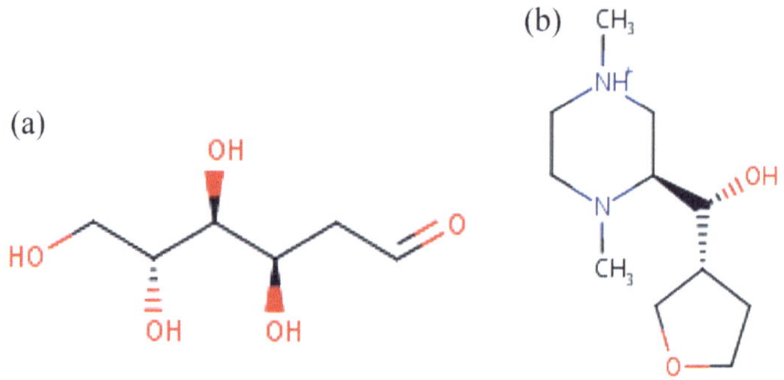

Fig. 5 The structures of **a** 2DG and **b** Chitin

Table 1 The ADME and toxicity analyses of 2DG and chitin

Compound	Similarity Score	Toxicity	ADME			
			Absorption (%)	Distribution (log L/Kg)	Metabolism	Excretion
2DG	N/A	Toxicity Class: II (Toxic)	86.433	0.916	No	No
Chitin	0.80	Toxicity Class: VI (Non-toxic)	86.983	0.914	No	No

of both 2DG and chitin were used to obtain their respective ADME values. Chitin met the ADME analysis benchmarks (Table 1).

6.1.3 Toxicity Test

In a drug design process, predicting compound toxicity is essential to identify non-toxic candidates. In this study, we used ProTox-II (http://www.tox.charite.de/protox_II/) to assess the non-toxicity of chitin. The SDF format of chitin and 2DG were converted into a molecular data file (MDL Molfile) using Open Bable software. ProTox-II classifies toxicity into six levels, from, Class-I to Class-VI, based on a range of lethal dose (LD50) values, where Class-I indicates a toxic compound, while Class-VI determines a non-toxic compound. Chitin was selected as a non-toxic compound whereas 2DG was determined as toxic (Table 1).

6.2 Structure-Based Screening

Structure-based screening is a method to predict the best interactions between ligands and target proteins. Ligands are ranked according to the binding affinity, where the highest binding affinity was taken as a promising potential drug molecule.

6.2.1 Human Hexokinase II Protein Preparation

In this study, the crystal structure of human HKII has been utilized as a drug target. HKII is a homodimer protein with highly similar N and C-terminal domains that together form half of the protein (Fig. 6). The crystal structure of HKII (2NZT.pdb) in protein data bank (PDB) file format (PDB ID: 2NZT) was retrieved from the Research Collaboratory for Structural Bioinformatics (RCSB) website (https://www.rcsb.org/). The native structure of the protein was complexed with a heteroatom, water molecule as well as unwanted ligand, which are crucial to be removed to make good protein–ligand interaction. These issues have been solved using AutoDock Tools (ADT) combined with the Python Molecular Viewer (PMV) tool. The 2nzt.pdb file was loaded on the ADT tool by choosing the 'file' option, where bonded atoms were represented as lines and non-bonded atoms such as metal ions, oxygen atoms and water molecules were shown as small squares. All unwanted molecules were removed, while all hydrogen atoms were added to the protein's PDB file by choosing the 'Add hydrogen' option of ADT tools. Furthermore, Kollman charges and compute Gasteiger charges were added to the crystal structure of the protein. These charges are significant to incorporate partial charges of the ligand that will be docked and to calibrate the scoring function. The prepared protein was saved in PDBQT file format, which is the extended form of PDB that includes all partial charges (Q) and atom types (T).

The water molecule might deform the binding site residues as well as the interactions between protein–ligand atoms. On the other hand, the heteroatom and unknown ligand have to be removed due to docking of the selected ligand and to ensure a smooth molecular docking process. Structural visualization was conducted using PyMOL software.

6.2.2 Chitin Preparation

The next procedure is preparing chitin before conducting molecular docking. The first step was to extract the PDB file of chitin from ZINC database. Kollman charges and computes Gasteiger charges were added by selecting the 'Add Charge' option on ADT tools. Additionally, the rotatable bonds of chitin and maximum torsional degree of freedom (TORSDOF) were set by choosing the 'Ligand' option and selecting 'Torsion Tree' and 'Detect Root', respectively. Torsion angle or torsion tree selection was required because the excess number of the rotatable bond increases the difficulty

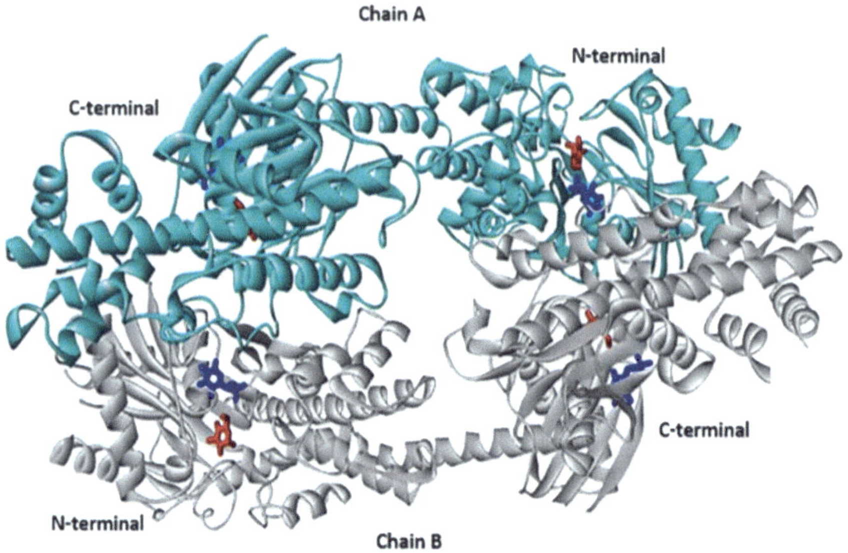

Fig. 6 The crystal structure of human HKII (Retrieved from PDB ID: 2NZT)

of the docking method. Finally, the prepared chitin was saved in the PDBQT file since AutoDock Vina only recognizes the PDBQT file format.

6.2.3 Molecular Docking of Chitin into HKII Using Auto Dock Vina

Molecular docking is commonly understood as protein–ligand binding. The goal of docking is to understand molecular recognition, favorable binding pose, and binding affinity of the ligand toward the 3D structure of the receptor protein. In this experiment, AutoDock Vina software was used that enhances the average accuracy of the binding mode predictions by calculating the scoring function. The scoring function is a mathematical tool employed to predict binding affinity between protein and ligand during docking processes. The general formula used for design scoring is as shown in Eq. 1.

$$c = \sum_{i<j} f_{t_i t_j}(r_{ij}) \tag{1}$$

where i has defined the atom or atomic type and f_{titj} represents atomic interaction, while rij defined the atomic distance (Trott and Olson 2010). The known ligand 2DG was removed for a re-docking purpose, where the active site of the protein structure is typically being validated by re-docking experiments. The ligand 2DG was extracted from native protein–ligand complexes to evaluate the conformation and, subsequently, docked back into the protein to validate the docking location.

Table 2 The binding affinity and H bond of 2DG and chitin

Compound	Binding Energy		Hydrogen bond number	
	Chain A	Chain B	Chain A	Chain B
	Terminal N, C	Terminal N, C	Terminal N, C	Terminal N, C
2DG	(−6.9) (−5.8)	(−5.9)(−6.0)	13, 9	11, 10
Chitin	(−6.3) (−6.1)	(−6.3)(−5.9)	7,7	8, 4

In this study, we analyzed the native structure of HKII using PyMOL and identified a cleft within both the N-terminal and C-terminal domain of the protein. The chitin compound and their query molecule 2DG were docked sequentially into this cleft. Notably, only the N-terminal domain was selected to conduct the docking process, as it plays a key role in regulating the stability of the enzyme, unlike the C-terminal domain. However, the catalytic residue of human hexokinase II has not yet been publicly available, making it a challenge to find the active site residues of HKII. To address this, we investigated the catalytic residues surrounding the cleft of HKII, where the known ligands (GLC and BG6) were bound using MGLTools (version 1.5.7), which were further validated by re-docking processes. In this work, AutoDock Vina program significantly enhanced the average accuracy of the binding mode predictions. In the Vina script, all information must be included such as grid box size of 2DG (Box size: $26 \times 24 \times 22$ A°), grid box center (X $= 43.567$, Y $= 35.439$, Z $= 36.27$), and measurement of space (1.000), ensuring the defined grid box encompasses the protein binding site to do sequential docking of chitin. The PDBQT format of the HKII protein, and all grid box information were occupied using MGLTools, while the SDF files of chitin were converted to a PDBQT file using OpenBable software. AutoDock Vina takes 10–15 min to complete the docking process (Tanbin et al. 2020).

Chitin exhibits binding energy ranging from −6.3 kcal/mol to −5.9 kcal/mol on both terminal ends of chain A and B, whereas 2DG itself poses binding energy −6.9 kcal/mol to −6.0 on both chain A and B (Table 2). Interestingly, the query molecule 2DG demonstrates good binding affinity (−6.9 kcal/mol) with 13 hydrogen bonds from which eight H bonds were formed with Gly233, Lys173, Ser155, Glu294, Thr172, Asn208, Asp209 and three halogen bonds formed with Phe156 and Ile229 (Fig. 7a). However, Chitin poses good binding energy with 7 hydrogen bonds, four of which are formed with Ser449, Asp84, and Arg91 and three salt bridge H bonds are formed with Asp84, Asp209, and Asp413 (Fig. 7b).

6.3 Molecular Dynamics Simulation

Molecular dynamics simulations are commonly employed to examine the behavior of molecules at the atomic level, providing insights into structure and interactions between protein and ligand molecules over a specified time period. In this study,

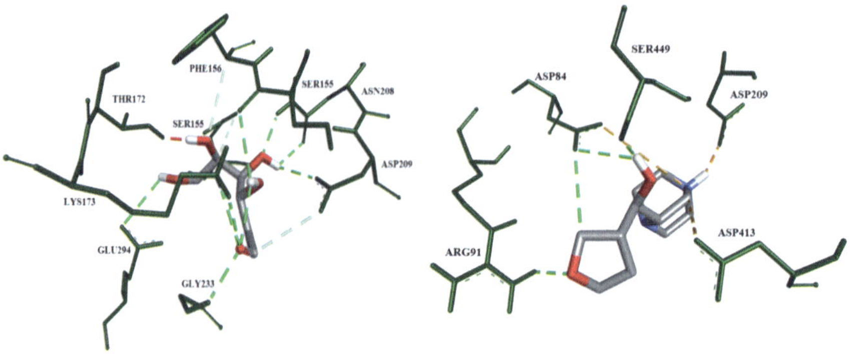

Fig.7 The dock poses of **a** 2DG and **b** Chitin

GROMACS version 5 was utilized to set up the molecular dynamics simulation process for the protein–ligand complex, with the simulation environment configured on a Linux operating system. Throughout the simulation process, GROMACS version 5, Xmgrace, and VMD were the primary tools used. The general workflow of MD simulation is shown in Fig. 8.

In this experiment, the HKII-2DG and HKII-Chitin complexes (referred to as HKII-ligand/compound) were used, which were generated from a previous molecular docking experiment. Before initiating the simulation, it was recommended to

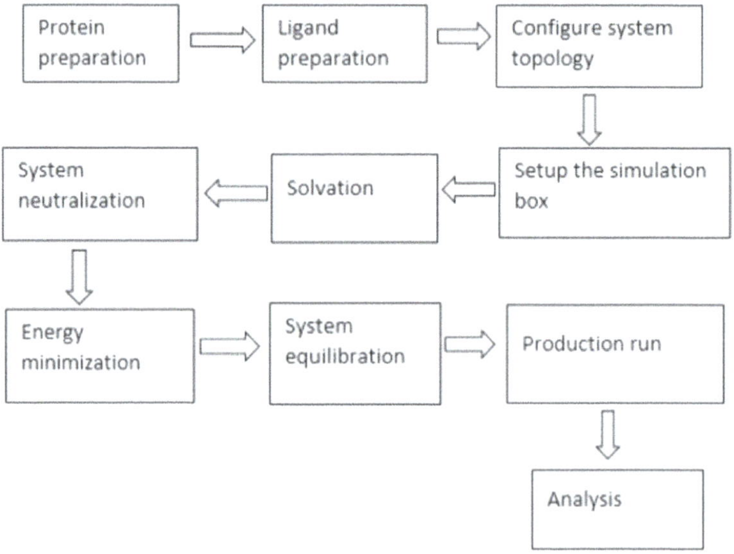

Fig. 8 The general workflow of MD simulation

organize all input and output files generated by GROMACS in the same directory. Since GROMACS does not recognize the 'complex. pdbqt' files produced from docking, these files were converted to pdb format for compatibility. The GROMOS96 54a7 force field was applied to define the potential energy, which governs the atomic interactions. The same force field was used during protein preparation. The Automated Topology Builder (ATB) server (https://atb.uq.edu.au) was employed to generate ligand topology files compatible with GROMOS96 54a7 force field. ATB is a tool where users can submit ligand structures to create the necessary topology files. After completing these preparations, the protein was ready for simulation. The production run began using 'pr.mdp' file, which included the simulation length; nsteps and dt parameters.

$$\text{Simulation time (ns)} = \text{nsteps} \times \text{dt}.$$

For atomic MD simulation, the dt is typically set to 0.002 (or 2 fs), and nsteps adjusted accordingly to achieve the desired simulation length. In this case, the simulation time was set to 20 ns, and the nsteps set to 10,000,000 for the production run. Once the simulation was completed, the trajectory was visualized using VMD, and analyses such as RMSD, protein ligand distance throughout the simulation, and formation of the hydrogen bond were carried out using Xmgrace software.

Both the HKII-Chitin (comp 1) and HKII-2DG complexes were equilibrated throughout the simulation, with minimal deviations in their protein backbones after 15 ns of simulation. Specifically, the backbone of the HKII-chitin (comp 1) complex showed a slight fluctuation compared to the reference HKII- 2DG complex toward the end of the simulation, with the RMSD values stabilizing at ± 0.5 Å. This suggests that the interaction between HKII with chitin (comp 1) is comparable to that of HKII and 2DG. To assess the binding topology of the ligands, snapshots of the HKII-2DG and HKII-chitin (compound 1) complexes were taken from the production simulation and superimposed (Fig. 9). The structures of both ligands were found to be coordinated similarly, indicating comparable levels of inhibition toward the enzyme. The binding interactions of the HKII-chitin (compound 1) and HKII-2DG complexes were carefully monitored and differentiated throughout the simulation trajectory. For root mean square fluctuation (RMSF), the fluctuations of each amino acid in the HKII complex with chitin followed a similar pattern to those observed for the 2DG reference compound. The stability of the protein-ligand complexes were assessed by calculating the H bond profiles using the GROMACS g_hbond tool. The analysis revealed that the HKII-2DG complex formed an average of six hydrogen bonds higher than the HKII-chitin complex (three hydrogen bonds) over the course of the simulation (Fig. 9). Figure 9 shows the trajectory of HKII-2DG and HKII-chitin (comp 1) extracted at 20 ns, where chitin (comp 1) binds to HKII via three H bond with the catalytic residues of Asp84, Asp209, and Asp413. In contrast, the reference compound 2DG forms six H bonds with residues Asp84, Asn89, Asp209, Thr232, Thr88, and Ser415 (Fig. 9) (Tanbin et al. 2020). This analysis indicates that 2DG exhibits stronger interactions with HKII under dynamic conditions during the MD simulation. While the number

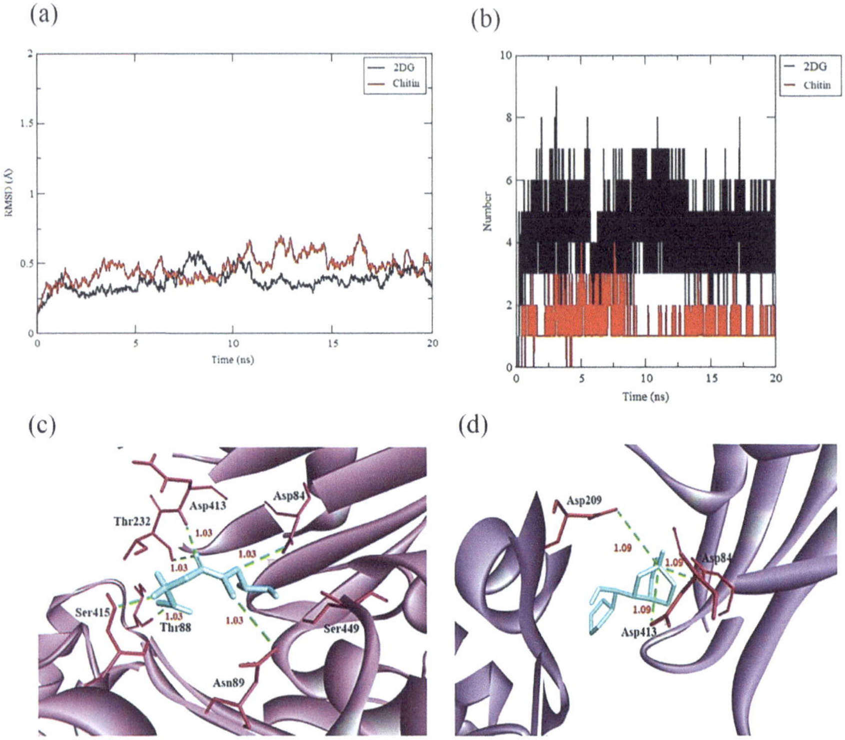

Fig. 9 The MD analysis of HKII-2DG and HKII-chitin (compound 1) complexes. **a** The Root Mean Square Deviation (RMSD) of HKII-2DG and HKII-chitin complex **b** The number of H bonds for both HKII-2DG and HKII-chitin complexes **c** The dock poses captured at 15 ns for HKII-2DG complex **d** The dock poses captured at 15 ns for HKII-chitin complex. (Tanbin et al. 2020)

of H bonds formed by chitin (compound 1) is lower than 2DG, it consistently maintains stable binding within the active site of HKII, particularly near the enzyme's active site.

7 Conclusion

Synthetic biology is revolutionizing the biotechnology industry, increasingly being applied in previously untapped markets. In this context, we emphasize the importance of synthetic biology to the bioeconomy, focusing on two of its defining features; its approach to research and development (R&D) and its carefully orchestrated, stakeholder-inclusive, community-driven model for sector growth. A relatively new field, synthetic biology emerged from the convergence of engineering, information technology, and biology. Since its inception, the term 'synthetic biology' has evolved into a broad concept, characterized by a framework that seeks to produce valuable biological products through the rational design of biological systems. This approach replaces the trial-and-error practices typical of conventional biotechnology with engineering principles at the heart of the R&D cycle. As a result, the sector becomes more stable, scalable, and adaptable, with the potential of automation and the ability to transition to carbon–neutral, efficient production methods for a wide variety of products. This chapter discusses the potential application of chitin as a drug and illustrates how synthetic biology, through in silico analyses, is shedding light on naturally occurring compounds for pharmaceutical development.

References

Agu, P. C., Afiukwa, C. A., Orji, O. U., Ezeh, E. M., Ofoke, I. H., Ogbu, C. O., Ugwuja, E. I., & Aja, P. M. (2023). Molecular docking as a tool for the discovery of molecular targets of nutraceuticals in diseases management. *Scientific Reports, 13*(1), 13398. https://doi.org/10.1038/S41598-023-40160-2

Ajjarapu, S. M., Tiwari, A., Ramteke, P. W., Singh, D. B., & Kumar, S. (2021). Ligand-based drug designing. *Bioinformatics: Methods and Applications*, 233–252. https://doi.org/10.1016/B978-0-323-89775-4.00018-3

Aranaz, I., Acosta, N., Civera, C., Elorza, B., Mingo, J., Castro, C., de los Llanos Gandía, M., Caballero, A. H. (2018) Cosmetics and Cosmeceutical Applications of Chitin Chitosan and Their Derivatives. *Polymers, 10(2),* 213. https://doi.org/10.3390/polym10020213

Azimi, B., Ricci, C., Fusco, A., Zavagna, L., Linari, S., Donnarumma G., Hadrich, A., Cinelli, P., Coltelli, M., Danti, S., Lazzeri, A. (2021). Electrosprayed Shrimp and Mushroom Nanochitins on Cellulose Tissue for Skin Contact Application. *Molecules, 26(14),* 4374. https://doi.org/10.3390/molecules26144374

Batool, M., Ahmad, B., & Choi, S. (2019). A structure-based drug discovery paradigm. *International Journal of Molecular Sciences, 20*(11). https://doi.org/10.3390/IJMS20112783

Bays, H. E., Evans, J. L., Maki, K. C., Evans, M., Maquet, V., Cooper, R., & Anderson, J. W. (2013). Chitin-glucan fiber effects on oxidized low-density lipoprotein: A randomized controlled trial. *European Journal of Clinical Nutrition, 67*(1), 2–7. https://doi.org/10.1038/ejcn.2012.121

Biala, G., Kedzierska, E., Kruk-Slomka, M., Orzelska-Gorka, J., Hmaidan, S., Skrok, A., Kaminski, J., Havrankova, E., Nadaska, D., & Malik, I. (2023). Research in the field of drug design and development. *Pharmaceuticals 16*(9), 1283. https://doi.org/10.3390/PH16091283

Boot, R. G., Renkema, G. H., Verhoek, M., Strijland, A., Bliek, J., de Meulemeester, T. M., Mannens, M. M., & Aerts, J. M. (1998). The human chitotriosidase gene. Nature of inherited enzyme deficiency. *Journal of Biological Chemistry, 273*(40), 25680–25685. https://doi.org/10.1074/jbc.273.40.25680

Chen, J., Luo, X., Qiu, H., Mackey, V., Sun, L., & Ouyang, X. (2018). Drug discovery and drug marketing with the critical roles of modern administration. *American Journal of Translational Research, 10*(12), 4302. /pmc/articles/PMC6325519/

Choi, J. P., Lee, S. M., Choi, H. I., Kim, M. H., Jeon, S. G., Jang, M. H., Jee, Y. K., Yang, S., Cho, Y. J., & Kim, Y. K. (2016). House dust mite-derived chitin enhances Th2 cell response to inhaled allergens, mainly via a TNF-α-Dependent pathway. *Allergy Asthma Immunol Res, 8*(4), 362–374. https://doi.org/10.4168/aair.2016.8.4.362

Drug Discovery World. (2016). With synthetic biology, drug discovery is going virtual. *Drug Discovery World (DDW).* https://www.ddw-online.com/with-synthetic-biology-drug-discovery-is-going-virtual-1422-201608/

Deore, A. B., Dhumane, J. R., Wagh, R., & Sonawane, R. (2019). The stages of drug discovery and development process. *Asian Journal of Pharmaceutical Research and Development*, *7*(6), 62–67. https://doi.org/10.22270/AJPRD.V7I6.616

Desai, N., Rana, D., Salave, S., Gupta, R., Patel, P., Karunakaran, B., Sharma, A., Giri, J., Benival, D., Kommineni, N. (2023). Chitosan: A Potential Biopolymer in Drug Delivery and Biomedical Applications. *Pharmaceutics, 15(4),* 1313. https://doi.org/10.3390/pharmaceutics15041313

Dhakal, A., McKay, C., Tanner, J. J., & Cheng, J. (2022). Artificial intelligence in the prediction of protein–ligand interactions: Recent advances and future directions. *Briefings in Bioinformatics, 23*(1), 1–23. https://doi.org/10.1093/BIB/BBAB476

European Community. (2014). Opnion on Synthetic Biology I. *Scientific Committees, European commission.* https://doi.org/10.2772/76553

Food and Drug Administration. (2018a). *Step 3: Clinical Research | FDA.* https://www.fda.gov/patients/drug-development-process/step-3-clinical-research

Food and Drug Administration. (2018b). *Step 1: Discovery and Development | FDA.* https://www.fda.gov/patients/drug-development-process/step-1-discovery-and-development

Food and Drug Administration. (2022). *Investigational New Drug Applications (INDs) for CBER-Regulated Products | FDA.* https://www.fda.gov/vaccines-blood-biologics/development-approval-process-cber/investigational-new-drug-applications-inds-cber-regulated-products

Garcia, S. N., Guedes, R. C., & Marques, M. M. (2019). Unlocking the potential of HK2 in cancer metabolism and therapeutics. *Current Medicinal Chemistry, 26*(41), 7285–7322. https://doi.org/10.2174/0929867326666181213092652

He, H., Sun, C., Weng, Y., Huang, H., Ni, P., Fang, Y., Xu, R., Wang, Z., & Liu, H. (2022). Catechol modification of non-woven chitosan gauze for enhanced hemostatic efficacy. *Carbohydrate Polymers, 286*, 119319. https://doi.org/10.1016/j.carbpol.2022.119319

Hughes, J. P., Rees, S. S., Kalindjian, S. B., & Philpott, K. L. (2011). Principles of early drug discovery. *British Journal of Pharmacology, 162*(6), 1239. https://doi.org/10.1111/J.1476-5381.2010.01127.X

Jayakumar, R., Chennazi, K. P., Srinivasan, S., Nair, S.V., Furuike, T., & Tamura, H. (2011). Chitin Scaffolds in tissue engineering. *International Journal of Molecular Sciences, 12*(3), 1876–1887. https://doi.org/10.3390/ijms12031876

Joshi, J., & Hanson, A. D. (2024). A pilot oral history of plant synthetic biology. *Plant Physiology.* https://doi.org/10.1093/PLPHYS/KIAD585

Jung, S., Vatheuer, H., & Czodrowski, P. (2023). VSFlow: An open-source ligand-based virtual screening tool. *Journal of Cheminformatics, 15*(1), 1–10. https://doi.org/10.1186/S13321-023-00703-1/FIGURES/10

Kandi, V., & Vadakedath, S. (2023). Clinical trials and clinical research: A comprehensive review. *Cureus, 15*(2). https://doi.org/10.7759/CUREUS.35077

Karki, N., Verma, N., Trozzi, F., Tao, P., Kraka, E., & Zoltowski, B. (2021). SSnet: A deep learning approach for protein-ligand interaction prediction. *International Journal of Molecular Sciences, 22*(3), 1–16. https://doi.org/10.3390/IJMS22031392

Khosravi, A. R., & Erle, D. J. (2016). Chitin-induced airway epithelial cell innate immune responses are inhibited by carvacrol/thymol. *PLOS ONE, 11*(7), e0159459. https://doi.org/10.1371/journal.pone.0159459

Khrunyk, Y., Lach, S., Petrenko, I., & Ehrlich, H. (2020). Progress in Modern Marine Biomaterials Research. *Marine Drugs,* 18(12), 589. https://doi.org/10.3390/md18120589

Kimber, T. B., Chen, Y., & Volkamer, A. (2021). Deep learning in virtual screening: Recent applications and developments. *International Journal of Molecular Sciences*, *22*(9). https://doi.org/10.3390/IJMS22094435

Kitney, R., & Freemont, P. (2012). Synthetic biology—The state of play. *FEBS Letters, 586*(15), 2029–2036. https://doi.org/10.1016/j.febslet.2012.06.002

Kulka, K., Sionkowska, A. (2023). Chitosan Based Materials in Cosmetic Applications: A Review. *Molecules, 28(4),* 1817. https://doi.org/10.3390/molecules28041817

Kumar, P. T. S., Srinivasan, S., Lakshmanan, V., Tamura, H., Nair, S. V., & Jayakumar, R. (2011). β-Chitin hydrogel/nano hydroxyapatite composite scaffolds for tissue engineering applications. *Carbohydrate Polymers, 85*(3), 584–591. https://doi.org/10.1016/j.carbpol.2011.03.018

Lim, W. A., Alvania, R., & Marshall, W. F. (2012). Cell biology 2.0. *Trends in Cell Biology, 22*(12), 611–612. https://doi.org/10.1016/j.tcb.2012.10.004

Li, Z., Liu, C., Hong, S., Lian, H., Mei, C., Lee, J., Wu, Q., Hubbe, M. A., & Li, M. (2022). Recent advances in extraction and processing of chitin using deep eutectic solvents. *Chemical Engineering Journal, 446,* 136953. https://doi.org/10.1016/j.cej.2022.136953

Liu, L., Zhang, X., Liu, Y., Zhang, L., Zheng, J., Wang, J., Hansbro, P. M., Wang, l., Wang, G., & Hsu, A. C. (2019). Chitinase-like protein YKL-40 correlates with inflammatory phenotypes anti-asthma responsiveness and future exacerbations. *Respiratory Research, 20*(1). https://doi.org/10.1186/s12931-019-1051-9

Maia, E. H. B., Assis, L. C., de Oliveira, T. A., da Silva, A. M., & Taranto, A. G. (2020). Structure-based virtual screening: From classical to artificial intelligence. *Frontiers in Chemistry, 8*, 481382. https://doi.org/10.3389/FCHEM.2020.00343/BIBTEX

Mazur, M., Olczak, J., Olejniczak, S., Koralewski, R., Czestkowski, W., Jedrzejczak, A., Golab, J., Dzwonek, K., Dymek, B., Sklepkiewicz, P. L., Zagozdzon, A., Noonan, T., Mahboubi, K., Conway, B., Sheeler, R., Beckett, P., Hungerford, W. M., Podjarny, A., Mitschler, A., Cousido-Siah, A., Fadel, F., & Golebiowski, A. (2018). Targeting acidic mammalian chitinase is effective in animal model of asthma. *Journal of Medicinal Chemistry, 61*(3), 695–710. https://doi.org/10.1021/acs.jmedchem.7b01051

Meng, X.-Y., Zhang, H.-X., Mezei, M., & Cui, M. (2011). Molecular docking: A powerful approach for structure-based drug discovery. *Current Computer-Aided Drug Design, 7*(2), 146. https://doi.org/10.2174/157340911795677602

Moussian, B., Yang, Q., & Fukamizo, T. (2019). Targeting chitin-containing organisms chitin: structure chemistry and biology. *Advances in Experimental Medicine and Biology, 1142*, 5–18. https://doi.org/10.1007/978-981-13-7318-3_2

Pertrenko, I., Bazhenov, V. V., Galli, R., Wysokowski, M., Fromont, J., Schupp, P. J., Stelling, A. L., Niederschlag, E., Stöker, H., Kutsova, V. Z., Jesionowski, T., & Ehrlich, H. (2017). Chitin of poriferan origin and the bioelectrometallurgy of copper/copper oxide. *International Journal of Biological Macromolecules, 104*, 1626–1632. https://doi.org/10.1016/j.ijbiomac.2017.01.084

Philp, J., & Winickoff, D. E. (2018). Realising the circular bioeconomy. *OECD Science, Technology and Industry Policy Papers.* OECD Publishing, 60. https://doi.org/10.1787/31bb2345-en

Pinzi, L., & Rastelli, G. (2019). Molecular docking: Shifting paradigms in drug discovery. *International Journal of Molecular Sciences, 20*(18). https://doi.org/10.3390/IJMS20184331

Qin, Z., & Zhao, L. (2019). The history of chito/chitin oligosaccharides and its monomer. The history of chito/chitin oligosaccharides and its monomer. In: Zhao, L. (Eds.), *Oligosaccharides of Chitin and Chitosan,* (pp. 3–14). Springer Singapore. https://doi.org/10.1007/978-981-13-9402-7_1

Qin, Y., Li, P. (2020). Antimicrobial Chitosan Conjugates: Current Synthetic Strategies and Potential Applications. *International Journal of Molecular Sciences, 21(2),* 499. https://doi.org/10.3390/ijms21020499

Roche, D. B., & McGuffin, L. J. (2016). In silico identification and characterization of protein-ligand binding sites Book or Report Section. *Methods in Molecular Biology.*https://doi.org/10.1007/978-1-4939-3569

Shin, W. H., & Kihara, D. (2018). Virtual ligand screening using PL-PatchSurfer2, a molecular surface-based protein-ligand docking method. *Methods in Molecular Biology (Clifton, N.J.), 1762*, 105–121. https://doi.org/10.1007/978-1-4939-7756-7_7

Singh, D., & Singh, B. (2020). Investigating the impact of data normalization on classification performance. *Applied Soft Computing, 97*, 105524. https://doi.org/10.1016/J.ASOC.2019.105524

Singh, N., Vayer, P., Tanwar, S., Poyet, J.-L., Tsaioun, K., & Villoutreix, B. O. (2023). Drug discovery and development: Introduction to the general public and patient groups. *Frontiers in Drug Discovery, 3*, 1201419. https://doi.org/10.3389/FDDSV.2023.1201419

Song, S., Zhao, Y., Yuan, X., & Zhang, J. (2019). β-Chitin nanofiber hydrogel as a scaffold to in situ fabricate monodispersed ultra-small silver nanoparticles. *Colloids and Surfaces A: Physicochemical and Engineering Aspects, 574*, 36–43. https://doi.org/10.1016/j.colsurfa.2019.04.047

Tanbin, S., Fuad, F. A. A., & Hamid, A. A. A. (2020). Virtual screening for potential inhibitors of human Hexokinase II for the development of anti-dengue therapeutics. *BioTech, 10*(1),https://doi.org/10.3390/biotech10010001

Tang, H., Fang, Z., Sun, Y., Li, B., Shi, Z., Chen, J., Zhang, T., & Xiu, Q., (2010). YKL-40 in asthmatic patients and its correlations with exacerbation eosinophils and immunoglobulin E. *European Respiratory Journal, 35*(4), 757–760. https://doi.org/10.1183/09031936.00034409

Theerawattanawit, C., Phaiyarin, P., Wanichwecharungruang, S., Noppakun, N., Asawanonda, P., Kumtornrut, C. (2021). The Efficacy and Safety of Chitosan on Facial Skin Sebum. *Skin Pharmacology and Physiology, 35(1)*, 23–30. https://doi.org/10.1159/000517965

Trott, O., & Olson, A. J. (2010). AutoDock Vina: Improving the speed and accuracy of docking with a new scoring function efficient optimization and multithreading. *Journal of Computational Chemistry, 31*(2), 455–461. https://doi.org/10.1002/jcc.v31:210.1002/jcc.21334

Tsurkan, M. V., Voronkina, A., Khrunyk, Y., Wysokowski, M., Pertrenko, I., & Ehrlich, H. (2021). Progress in chitin analytics. *Carbohydrate Polymers, 252*, 117204. https://doi.org/10.1016/j.carbpol.2020.117204

Wang, X., Song, K., Li, L., & Chen, L. (2018). Structure-based drug design strategies and challenges. *Current Topics in Medicinal Chemistry, 18*(12), 998–1006. https://doi.org/10.2174/1568026618666180813152921

Wang, Y., Jiao, Q., Wang, J., Cai, X., Zhao, W., & Cui, X. (2023). Prediction of protein-ligand binding affinity with deep learning. *Computational and Structural Biotechnology Journal, 21*, 5796–5806. https://doi.org/10.1016/J.CSBJ.2023.11.009

World Economic Forum. (2016). These are the top 10 emerging technologies of 2016. *Emerging Technologies, Annual Meeting of the New Champions*. https://www.weforum.org/stories/2016/06/top-10-emerging-technologies-2016/

Woodrow Wilson International Center for Scholars (WWICS). (2013). *Tracking the Growth of Synthetic Biology: Findings for 2013.*

Wu, Z., Li, W., Liu, G., & Tang, Y. (2018). Network-based methods for prediction of drug-target interactions. *Frontiers in Pharmacology, 9*. https://doi.org/10.3389/FPHAR.2018.01134

Wu, Z., Ma, H., Liu, Z., Zheng, L., Yu, Z., Cao, S., Fang, W., Wu, L., Li, W., Liu, G., Huang, J., & Tang, Y. (2022). WSDTNBI: A novel network-based inference method for virtual screening. *Chemical Science, 13*(4), 1060–1079. https://doi.org/10.1039/D1SC05613A

Xie, Y., Yang, Y., He, Y., Wang, X., Zhang, P., Li, H., & Liang, S. (2020). Synthetic biology speeds up drug target discovery. *Frontiers in Pharmacology, 11*, 119. https://doi.org/10.3389/fphar.2020.00119

Yang, C., Chen, E. A., & Zhang, Y. (2022). Protein–ligand docking in the machine-learning Era. *Molecules, 27*(14), 4568. https://doi.org/10.3390/MOLECULES27144568

Yu, P., Xie, J., Chen, Y., Jinming, L., Liu, Y., Bi, B., Luo, J., Li, S., Jiang, X., & Li, J. (2020). A thermo-sensitive injectable hydroxypropyl chitin hydrogel for sustained salmon calcitonin release with enhanced osteogenesis and hypocalcemic effects. *Journal of Materials Chemistry B, 8*(2), 270–281. https://doi.org/10.1039/C9TB02049G

Zhang, X., Wu, F., Yang, N., Zhan, X., Liao, J., Mai, S., & Huang, Z. (2021). In silico methods for identification of potential therapeutic targets. *Interdisciplinary Sciences: Computational Life Sciences, 14*(2), 285–310. https://doi.org/10.1007/S12539-021-00491-Y

Zhao, L., Zhu, Y., Wang, J., Wen, N., Wang, C., & Cheng, L. (2022). A brief review of protein–ligand interaction prediction. *Computational and Structural Biotechnology Journal, 20*, 2831–2838. https://doi.org/10.1016/J.CSBJ.2022.06.004

Collagen Production from Animal Husbandry for Circular Economy

Munira Shahbuddin, Husna Ahmad Tajuddin, Amnatasha Amerai Khan, and Dahlia Shahbuddin

Abstract Tissue regeneration in adult mammal does not occur spontaneously during wound healing but going through the process of remodelling and re-epidermisation of the tissue, resulting in contraction and scar formation. The extraction of collagen and fabrication of collagen scaffold are important to the outcome of the treatment. The specific and important chemical analogues and properties in collagen can be lost during the process of extraction and fabrication due to exposure to heat, moist and atmospheric condition. In this work, we looked into the details of extraction of collagen from bovine tendon, starting from the use of freeze drying to assist with swelling, the use of different concentrations of acetic acid to dialysis for 3–5 days and fabrication of collagen scaffold. From this work, we emphasis that the tiny details in extraction and fabrication process of collagen scaffold are critical to the effectiveness of the outcome of regenerative scaffold.

Keywords Collagen · Scaffold · Tissue engineering and circular economy

1 Introduction

Regenerative and sustainable bioeconomy is projected to generate 170 billion ringgit annually for Malaysia economy by 2020 under Bioeconomy Transformation Program. Malaysia's Biotechnology Economic Transformation cited Health care industry to be one of the core needs in the development of economy in the region. Along with the initiative for this endeavour which started in 2011, Malaysia is aimed to develop and master the bioeconomy in the region for wealth generation and well-being through works in agricultural, healthcare and industrial technology as well as in R&D and technology acquisition. Regenerative medicine and skin regeneration

M. Shahbuddin (✉) · H. A. Tajuddin · A. A. Khan
Department of Chemical Engineering and Sustainability, Kulliyyah of Engineering, International Islamic University of Malaysia (IIUM), Jalan Gombak, 53100 Kuala Lumpur, Malaysia
e-mail: munirashah@iium.edu.my

D. Shahbuddin
School of Biological Sciences, Universiti Sains Malaysia, 11700 Pulau Pinang, Malaysia

F. Adyani Ahmad Fuad et al. (eds.), *Circular Bioeconomy*,
https://doi.org/10.1007/978-981-97-7010-6_3

is a new field in Malaysia and has gained a lot of attention in Private and Public Health Sector as well as in Foreign Direct Investment (FDI) due to the country's economic growth and increased spending in healthcare. The invention and development of regenerative template for organ regeneration which have been successful for skin and nerve regeneration has improved the lives of many people around the world, especially those of burnt and accident victims, offered a new lease of life and improvement of life quality and health. It is important to note that the general public health sector is of importance in the development of a country and contributing directly to the national growth and Gross Domestic Produce (GDP).

Under the nine thrusts of Malaysia National Biotechnology Policy in the development of biotechnology for wealth creation and national well-being, thrust 4–9 mentioned of R&D technology acquisition, human capital development, financial infrastructure development, regulatory framework development, strategic development and government support and commitment. Therefore, collaborative effort and support from ASEAN member countries are much needed to tap the potential talents and ideas to meet these endeavours. ASEAN countries can work together to identify the specialties in specific sectors i.e. halal certified, organic farming and safe husbandry and biotechnological practices. With ASEAN Economic Cooperation (AEC), it is projected that Asia will be the second largest market for biomedical devices and healthcare after United States. Countries such as Thailand, Singapore, Malaysia and Indonesia are considered as hotspots for sales, investment and business opportunities. It is important for ASEAN to be able to produce their products to meet the demand of emerging market and create economic sustainability in the region.

Bioeconomy is the term that encompasses the production, utilisation and conservation of biological resources, including related knowledge, science, technology and innovation to provide information, products, processes and services across all economic sectors aiming towards a sustainable economy (San Juan et al. 2019). It is a major new industry that bioeconomy refers to all economic activities derived from the continued commercial applications of biotechnology that capitalises on biological resources and their conversion into domestic products, food, feed, chemicals, energy, pharmaceuticals and healthcare wellness products. The application of biotechnology has several main goals: adding value to the current bio-based production, reduce dependence on depletable natural resources, environmental sustainability and elevate the quality of life of the nation in its strive towards development. Malaysia is a nation rich in natural resources, and with this advantage, the pursuit for the comprehensive development of bioeconomy is not only logical, but also crucial in nation development and economy (Agochukwu et al. 2018). Towards this pursuit, many measures had been implemented targeting not only the elements of research and development but a complete value chain from the upstream to commercialisation.

Bioeconomy utilises the nation's comparative advantage in natural resources as well as providing an answer to limited food production, climate change and environmental sustainability. There are three stages of Bioeconomy in which circularity aspects are cross-cutting that includes biomass production, collection, processing and use and sustainable end-of-life option.

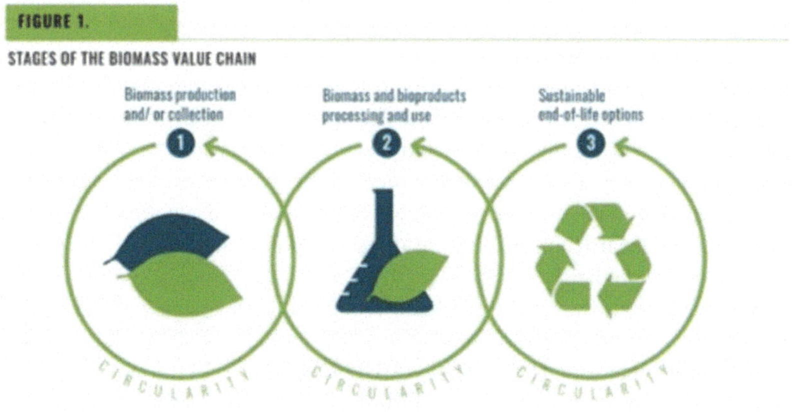

Fig. 1 A comprehensive approach to Circular Bioeconomy from resources to end-of-life

The concept of circular economy relies on the effective utilisation resources and waste (not only biological) in the economy cycle as long as possible before these resources reach the end-of-life stage. Circularity is the is the key aspect of sustainability where the central element lies in the adoption of 4R approach; Reduce, Reuse, Recycle and Recover where less inputs are used and less waste is produced. Responding to sustainability goals requires maximum utilisation of all raw materials produced and integration of activities throughout all production-to-consumption stages (Fig. 1).

In 2005, the Malaysian Government launched the National Biotechnology Policy (NBP) which recognised biotechnology as one of the key strategic drivers to propel the country's social and economic development in pursuit of the status of a developed nation (Loh, T. C. 2004). It aimed to further develop three economic sectors namely agriculture, healthcare and industrial manufacturing, as well as to support the growth of an enabling eco-system throughout the scientific, academic and business communities in the country. The key national development goals for NBP were set in three phases of a five year plan; Phase I for Capacity Building (2005–2010), Phase II on Science to Business (2011–2015) and Phase III to develop Global Business (2016–2020).

Agriculture is one of the sectors identified in the National Key Economic Areas (NKEAs) and is expected to grow by 3.5 per cent per annum, contributing 8.2 per cent to total GDP under the 11th Malaysia Plan, with the agro-food sub-sector growing to an expected 5.4% per annum. The sub-sector will be transformed into a high-income and sustainable industry through innovative R&D and modernisation initiatives. To meet this goal, Agensi Inovasi Malaysia (AIM) had developed an initiative called "National Biomass Strategy 2020" (NBS 2020) as a national blueprint on how Malaysia can leverage its available biomass across a diversified portfolio of higher value downstream uses to create new industries and jobs. NBS 2020 covers all types of biomass sources such as palm oil, forestry, rubber, dedicated crops, municipal waste, etc. across all sectors ranging from bioenergy, advanced biofuels and biochemical up to, eventually, end-products.

Bioeconomy is part of the 4th industrial revolution that changes the landscape for physical economy with the use of big data, 5G technology and artificial intelligent in decision making, innovation, creation and ways of doing businesses. The revolution may lead to changes in a way of which biotech-related activities such as developing organs for human transplantation, production of bio-fuel from genetically engineered algae, producing meat from cell culture or production of antibodies and medicine from personalised medicine.

Animal husbandry industry have been recognise as an important and integral component in agro-industry and bioeconomy that garner worldwide interest due to its high-income generation (Nigel et al., 2010). Globally, 1.3 billion people depend of livestock for their employment while billion other depend on livestock to produce living sustenance and comfortable means for their families. Europe and United States are the biggest producers of animal husbandry products in the twentieth century which now surpassed by China and India, where 80% increase was recorded in the meat sector and expected to rise in 2022 (FAOSTAT 2016).

Although Malaysia is being self-sufficient in poultry, the country is a net importer of ruminant-based meat products, particularly beef and mutton. The country is heavily reliant on imported feed-stock and resources constrain. Both beef and mutton produce recorded progressive growth during the period 1998–2000 with an average rate of 34 per cent and 19 per cent, respectively and despite this growth in meat production, beef and mutton still meet only 21 per cent of total annual domestic demand (Loh 2004).

Based on the report by Fadillah and Idrus, in the breakdown of bioeconomy contribution to GDP by subsectors in 2014, livestock contributed to 9% while another 9% contributed from a shared portion of fats and oils of vegetables and animals (Fadillah & Idrus, 2015). Although the data provided by the report was limited and lack of specific information in the identification and measurement of bioeconomy contribution to GDP, it is undoubted that Malaysia has to embark into a circular economy that utilises its1 resources effective from the source to the end-of-life and venture into territories of biotechnology and pharmaceutical industries.

2 Circular Economy in Animal Husbandry and Biotechnology

The bioeconomy essentially revolves around the production of standard goods and services utilising renewable feedstocks. The OECD's 2009 report; The Bioeconomy to 2030: Designing a Policy Agenda established some restrictions on the notion of the bioeconomy, but it has since grown significantly. It now includes a wide range of industries and activities, such as waste management, chemicals, food, agriculture, dairy and forestry. Biotechnology is only one aspect of the bioeconomy. In line with the idea of a circular economy, it is now viewed as a new method of production that will gradually replace fossil-based manufacturing (Philp and Winickoff, 2018). The manufacturing of high-value compounds from sustainable sources with less energy and waste is now possible, thanks to biotechnology, which makes use of enzymatic processes.

An early illustration of this would be the enzymes in laundry detergent, which enable effective cleaning at lower temperatures (i.e. with reduced energy use). Enzymatic catalysis is now being utilised to produce entire species, ranging from bacteria to yeast to plants. Complex and sought-after molecules can be created simply using glucose as a starting material, from the creation of insulin in genetically altered bacteria to the synthesis of artemisinic acid—a malaria medication precursor—in designed yeast. White biotechnology has benefited greatly from synthetic biology, which applies engineering ideas to biology and allows for the rational design or enhancement of production routes. White biotechnology is ideally positioned to serve as the cornerstone for a circular bioeconomy, with synthetic biology being hailed as one of the ten most significant emerging technologies by the World Economic Forum 2016 and biotechnology being named one of the six most important technologies for our sustainable future by the EU. The synthetic biology biodesign cycle and impacts on circular bioeconomy is shown in Fig. 2.

3 Synthetic Biology Speeds up Drug Target Discovery

Animal husbandry is at the centre of global sustainability challenges such as water scarcity, climate change and the rising of temperature, pollution and top soil degradation. Livestock sector is particularly sensitive to environmental problems and sustainability, and responsible to 40% of global greenhouse gases (GHG) emission worldwide. The severity of our climate situation has reached a tipping point where now consumers demand food producers and farmers to adhere to environmental friendly and sustainability approach.

To make livestock production in line with Sustainable Development Goals (SDGs) and climate mitigation target, a radical rethinking of animal husbandry practice model is required. In 2019 the European Union (EU) had placed animals and livestock farming at the top of public agenda, sectors ranging from animal health to feed, to breeding animal farming and farmers' practices.

In agro-industry, animal husbandry refers to the practice of animal reproduction and increment of livestock productivity and quality. There are two fundamental processes in animal husbandry, namely slaughter and processing in which both produce waste such as bones, carcasses, blood, organs such as eyes, brain and head, as well as skin. The waste from these processes was high of economic value for biopharmaceuticals and domestic products such as detergent, soap, cosmetics and chemicals.

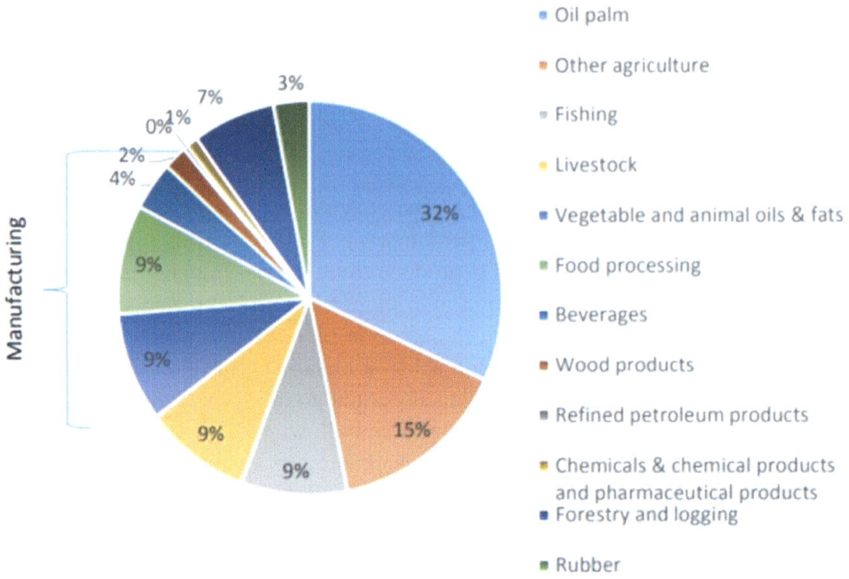

Fig. 2 Breakdown of bioeconomy contribution to GDP by subsectors in 2014, by value. Synthetic biology makes use of existing genetic modification technologies with the specific goal of accelerating and facilitating the process, which includes making it more predictable (European Community 2014). DNA-based circuits, synthetic metabolic pathway engineering, synthetic genomics, protocell assembly and xenobiology are some of the fields of study that fall under the umbrella term "synthetic biology"

Animal husbandry is one of the main sources for collagen manufacturing, where bovine sources accounted for a substantial volume share of nearly 40% in 2020 due to the availability of bovine-based raw material and lower price compared to porcine and marine. The market for hydrolysed collagen is anticipated to grow substantially due to increasing demand and use in tissue repair and dental applications.

Circular Economy (CE) is a popular theme in the sustainability field, where the waste will be fully utilise throughout the chain supply using effective waste management practice, environmental protection and conservation and technological innovation. From CE perspective, animal husbandry can offer a multitude of opportunities and wealth in comparison to the linear economy model which utilises resources and produces waste that can be damaging to the environment and health, with more than a quarter of produce ended up in landfill. In two models of CE implemented in China, animal husbandry wastes were exchanged as resources with other parts for fertilisation and land use.

4 Animal Husbandry and Biosafety

For years, animals have been used to produce metabolites, medicines, factors and hormones for human and laboratory use. Biopharmaceutical practices transform animals into "bioreactor" or "factory" to produce those elements in eggs, meat, milk and organs which later can be harvested for further use in research and development, as well as for medicine and therapeutic uses.

Examples of biotechnological use in animal husbandry include the creation of transgenic animals (animals with one or more genes introduced by human intervention), to produce genetically modified animals that able to synthesise proteins, improved growth rates and resistant to specific diseases. For decades, farmers have been improving livestock herds through enhanced animal husbandry practices such as cross-breeding, fortified ruminant and appropriate conditions of pen and shelter.

The coronavirus crisis demonstrates what is wrong with how the world feeds itself and reliance to "specialisation" economy, which brought into the question of sustainability and vulnerability of agro-animal husbandry. Specialisation limit and narrow production and economic activities to artificially create "economy advantages" at producing goods at lower costs and high efficiency to benefit customers, while neglecting the possibilities of new income generation through circular bioeconomy.

5 Venturing into High Income Biotechnology from Animal Husbandry

Among the polymers, collagen has shown particular interest because it is biodegradable, nontoxic, skin- and environmentally friendly ingredients can be obtained at low cost from food and forestry waste, respectively. The existing and potential uses of residues should always be included in the feasibility analysis of residue-based bioeconomy initiatives. For example in animal husbandry industry, collagen from bovine Achilles tendon can be made in tissue regenerative scaffold for wound and burn treatment, while serum can be extracted from the blood for biotechnology and medical laboratories uses in cell culture and therapeutics.

Malaysia need to venture in high technology R&D industry to become high-income generation country. Smithfield Foods, the world's largest pork producer originated from the United States of America, has established a separate bioscience unit to supplying pig parts for medical uses, with the ultimate goal of selling pig organs for transplantation into humans, known as xenotransplants. Smithfield is the subsidiary of China's WH Group had joined a public–private tissue engineering consortium funded by an $80 million grant from the U.S. Department of Defense. Smithfield is the only pork producer, joining healthcare including Abbott Laboratories, Medtronic and United Therapeutics Corp. Transplants are used for people diagnosed with organ failure and who have no other treatment options. Transplants from animals could help close a critical gap to help those in need. The United Network for Organ Sharing estimates that, on average, 22 people die each day while waiting for a transplant.

The U.S. market for pork by products used for medical, pet food and non-food purposes is valued at USD100 billion. The market is now going for biotechnology and medical industry with potential animal-to-human transplantation. Smithfield has deals in the works to supply pig organs for human transplantation (Steenhuysen et al., 2017). To date, each year Smithfield slaughtered 16 million hogs for its blood, skin, heart-valve, antibodies and other biological materials for medical and healthcare use. Engineering of highly precise and well-defined microarchitecture onto a biomaterial is often complex and tedious, and the scaffolds with these features are susceptible to surface and material disintegration during physical manipulation. Other attempts to modulate collagen scaffold topological features often face problems in the base material, thus preventing cell migration and blood supply. For biomedical applications, a detailed understanding of scaffolds and their interfaces is required for better tuning of biomaterials to suit the microenvironments. At present, there exists an unmet need of enabling the translation of microarchitectural principles into actual clinical applications.

6 Production of Collagen Scaffold with Well-Defined Microstructure from Bovine

The aim of this work is to investigate the geometrical correlation of collagen scaffold fabrication to physical and chemical structures using differential scanning calorimetry (DSC). Here, we present a novel view on these numerous control methods through the concept that each of these methods for controlling freeze casting by alteration of characteristic repulsive and attractive forces that govern the interactions between the solid loading particles and approaching freezing front during freezing of the slurry as well as the application of magnetic forces in the fabrication of anisotropic structure of the scaffold. This new system of understanding control over freeze casting can be applied to a wide variety of other fabrication and manufacturing techniques such as particle distribution in polymer-matrix composites and selective solidification of anisotropic metals for other applications from tissue engineering. In this work we examined the developments in both the design and utility of collagen scaffold along with their potential impact on future research and clinical application.

The idea of tissue engineering and regenerative medicine have garnered a lot of researcher's attention as it provides a goal to regenerate a fully functioning organs that can replaced original organs in human body. Through this idea the researchers have come up with a variety of technique and processes that can lead to successful regenerate tissue or organ. From the recent years, the research of tissue engineering and regenerative medicine has evolved when scaffolds are used as a tool to induce tissue regeneration. The trends of utilising scaffolds in tissue engineering and regenerative medicine have been on the rise seeing by the amount of journals and articles being published discussing the topics. Scaffold can provide structural and mechanical support for the cell by acting as a substrate for cellular growth, proliferation and inducing regeneration of injured tissue or organ. Due to its properties that match to native tissue structure, collagen-based scaffold is usually used as it encourages cellular binding, it has useful antigenic, haemostatic and mechanical (Laurent et al. 1987).

Properties of scaffolds are essential as it will determine the characteristics of tissue regenerated (Eltom et al. 2019). The criteria's needed by the scaffold are chemical identity of ligands, pore size and degradation rate of the scaffold. In order to manipulate the scaffold properties, the fabrication process play a very important role. There are various fabrication processes for collagen scaffold such as electrospinning, 3D printing and lyophilisation (Dong & Lv 2016). Different fabrication process will produce different structure of the scaffold. This is due to the type of equipment used in the process for example a fibrous scaffold can be produced by electrospinning. In this project, lyophilisation or freeze drying process was used to fabricate collagen scaffold. Three different dimension of will be used which are 15 cm × 10 cm × 3.5 cm for large aluminium tray, 8.5 cm × 8.5 cm × 4 cm for small aluminium tray and petri dish with diameter of 10 cm and height of 1 cm to observe whether it would affect the collagen scaffold structure and properties.

7 Materials and Methods

7.1 Fabrication of Collagen Scaffold

7.1.1 Extraction of Collagen from Bovine Legs

The experiment was started with the extraction process, by skinning the bovine legs to expose its tendon then; by using scalpel the tendon was cut to separate it from the legs. Figure 2 showed the exposed tendon from bovine legs. The tendon was cleaned and kept in the freezer. After the tendon being freezes, the tendons were immersed in 0.5 M acetic acid (CH_3COOH) for a week and then the mixture was blended in a blender. The mixture was poured into 50 ml centrifuge tube then, a scoop of sodium chloride (NaCl) was added into the mixture by using spatula. The tube was shaken until the mixture becomes homogeneous and was stored in refrigerator for 2 days at 4 °C to let the solution separate. Then, dialysis process was carried out on the collagen using dialysis membrane with molecular cut off ranging from 20 to 30 kDa and sodium phosphate dibasic solution with pH 7.2 was used as buffer for 7 days. Dialysis buffer was changed twice a day. Dialysis was done to further purify and separate collagen from other unwanted protein.

7.1.2 Freeze Drying of Collagen Scaffold

After extraction of collagen process, the collagen suspension was poured into a 15 cm × 10 cm × 3.5 cm stainless steel aluminium tray and then the suspension was put in a −20 °C freezer for one night. The collagen was freeze dried for 4 days to reach −80 °C temperature as the freezing rate for the freeze drier is −20 °C per day. After 4 days, the freeze dried collagen suspension will be in powder form. The powdered form of collagen was dissolved in 0.05 M of acetic acid with 5 mg/ml collagen concentration. The solution was chilled overnight. Next, 33 ml of collagen solution was poured into 15 cm × 10 cm × 3.5 cm stainless steel aluminium trays and is put in −20 °C freezer overnight. After that, the aluminium tray was put into pre-cooled freeze dryer. The temperature of the freeze dryer must be maintained at −80 °C at all times for 60 min as the freezing rate will influence the pore structure of the scaffold. The aluminium tray was put in freeze dryer for the next 4 days. These protocols were repeated with 8.5 cm × 8.5 cm × 4 cm stainless steel aluminium trays and plastic petri dish with dimension of 1.5 cm height and 10 cm diameter.

7.1.3 Crosslinking of Collagen Scaffold

Crosslinking process was to introduce covalent crosslink between polypeptide chains of the collagen fibre without denaturing the collagen into gelatine. After freeze drying, the collagen was in sheet form. The collagen sheet was carefully detached from the aluminium tray. For dehydrothermal treatment (DHT) process, the collagen sheet was put in aluminium foil wrapper and was put into oven of temperature 105 °C together with 50% of carbon dioxide, CO2 overnight. After overnight in the oven, the collagen sheet was taken out and cut into 1 cm × 1 cm small pieces. Each piece was put into 12 well plates. Each plate column was filled with control (0.05 M acetic acid), 0.1%, 0.5% and 1.0% of glutaraldehyde. The collagen sheet was left to be treated with glutaraldehyde overnight. After overnight of glutaraldehyde crosslinking, the collagen sheet was washed with distilled water 3 times before storing the sheet in 0.05 M of acetic acid until further use (Harley & Yannas 2013). The sheet was dried overnight in fume hood to remove any remaining acetic acid. Then the scaffold was analysed by using differential scanning calorimeter (DSC).

7.1.4 Analysis Using Differential Scanning Calorimetry

Firstly, the samples were labelled in the 12 well plates. Sample A was for control and B was for 1.0% of glutaraldehyde for petri dish. Large aluminium tray sample was labelled C for control and D was for 1.0% of glutaraldehyde. Small aluminium tray sample was labelled E for control and F for 1.0% of glutaraldehyde. The sample aluminium tray was weighed with the lid and then the samples were scraped from the well plate and put it in the sample aluminium tray. Then, again weight the sample aluminium tray that already filled with sample. The lid was placed onto the sample aluminium tray and the aluminium tray was placed in the differential scanning calorimeter. The parameters for DSC were temperature range, gas flow rate and type of gas used. The temperature range was set at 25–200 °C. Nitrogen gas with a flow rate of 50 ml/min was used to purge oxygen. DSC was used to analyse and to get data for the sample and graph of heat flow against temperature was plotted.

8 Result

8.1 The Effects of Freeze Drying, Dehyrdrothermal Treatment (DHT) and Glutaraldehyde Crosslinking on the Fabrication of Materials

From Fig. 3, the first endothermic peak observed in control and 1.0% GTA was between 38.12 °C and 107.06 °C; this is corresponding to dehydration process of surface water. The second endothermic phenomenon observed in the 1.0% GTA was in the temperature range between 132.96 °C and 179.71 °C. This was related to the loss of hydrogen bonds, so that the phenomenon of protein denaturation could be initiated in this interval in which the tertiary structure is lost. This phenomenon is reversible if the protein is newly hydrated. The collagen scaffold fabricated using with and without the treatment of glutaraldehyde had three distinct exothermal peaks in the thermogram, with Tg of 42 °C and total enthalpy amasses 121.69 $J.g^{-1}$ which is not so much different with the collagen scaffold that was treated with 1% of glutaraldehyde with 123.74 $J.g^{-1}$. The Tg of collagen scaffold treated with 1% glutaraldehyde is 52 °C. It is obvious that the large size of the aluminium tray reduces an even collagen concentration distribution throughout the sample, with weaker and thinner structure at the centre. The Tg of DHT treated and crosslinked collagen scaffold is 52 °C, and did not show much difference from the non-crosslinked collagen scaffold.

From Fig. 4, the first endothermic peak observed in control and 1.0% GTA was between 38.12 °C and 107.06 °C; this is corresponding to dehydration process of surface water. The second endothermic phenomenon observed in the 1.0% GTA was in the temperature range between 132.96 °C and 179.71 °C. This is related to the loss of hydrogen bonds, so that the phenomenon of protein denaturation could be initiated in this interval in which the tertiary structure is lost. This phenomenon is reversible if the protein is newly hydrated. From Fig. 4, collagen scaffold in the large aluminium tray with and without the treatment of glutaraldehyde had three distinct exothermal peaks in the thermogram, with Tg of 42 °C and total enthalpy amasses 121.69 $J.g^{-1}$ which is not so much different with the collagen scaffold that was treated with 1% of glutaraldehyde with 123.74 $J.g^{-1}$. The Tg of collagen scaffold treated with 1% glutaraldehyde is 52 °C. It is obvious that the large size of the aluminium tray reduces an even collagen concentration distribution throughout the sample, with weaker and thinner structure at the centre. The Tg of DHT treated and

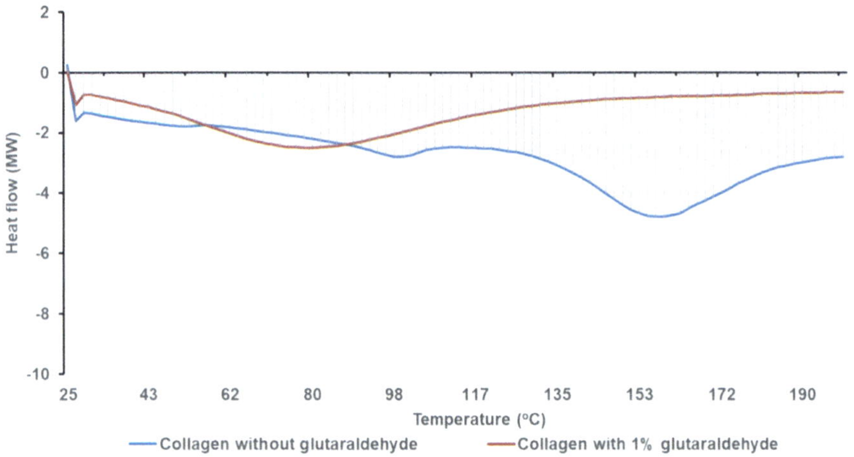

Fig. 3 DSC thermogram of collagen scaffold fabricated in 15 cm × 10 cm × 3.5 cm stainless steel aluminium tray without glutaraldehyde and with 1% glutaraldehyde heated from room temperature to 200 °C at 10 °C per minute

crosslinked collagen scaffold is 52 °C, and did not show much difference from the non-crosslinked collagen scaffold.

In Fig. 5, the first endothermic peak observed in control and 1.0% GTA was between 90.92 °C and 142.68 °C; this is corresponding to dehydration process of surface water. The second endothermic phenomenon observed in the 1.0% GTA was

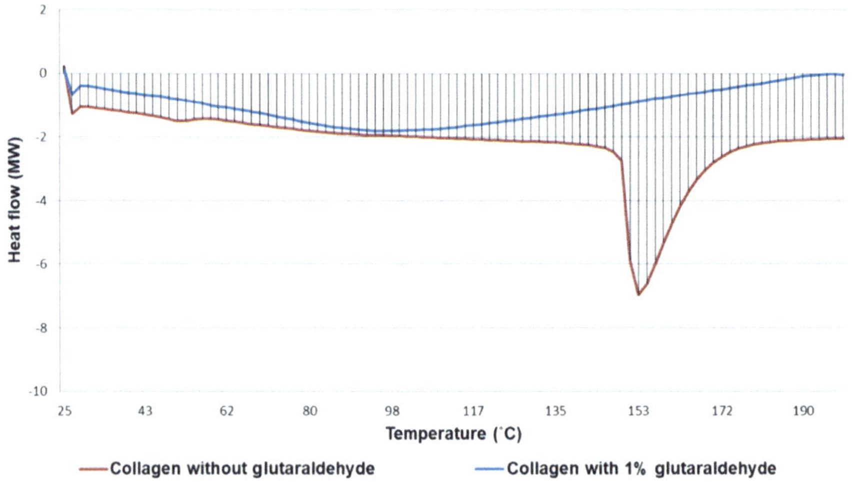

Fig. 4 DSC thermogram of collagen scaffold fabricated in 8.5 cm × 8.5 cm × 4 cm stainless steel aluminium tray without glutaraldehyde and with 1% glutaraldehyde heated from room temperature to 200 °C at 10 °C per minute

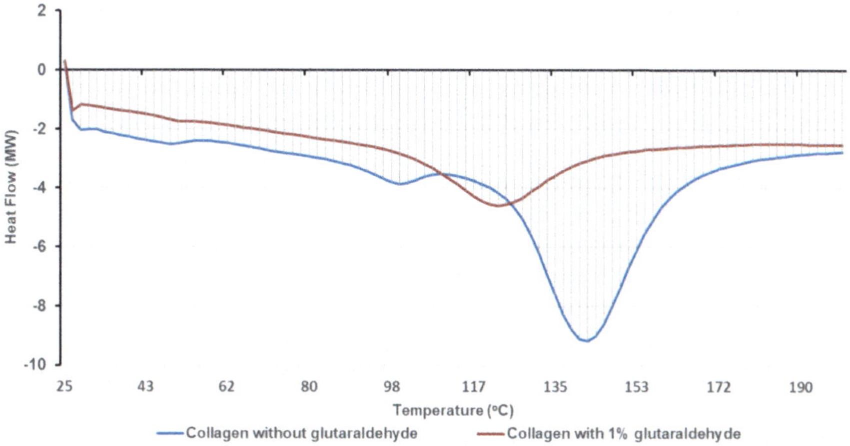

Fig. 5 DSC thermogram of collagen scaffold fabricated in 7.5 cm radius petri dish without glutaraldehyde and with 1% glutaraldehyde heated from room temperature to 200 °C at 10 °C per minute

in the temperature range between 124.31 °C and 160.86 °C. This was related to the loss of hydrogen bonds, so that the phenomenon of protein denaturation could be initiated in this interval in which the tertiary structure is lost. This phenomenon is reversible if the protein is newly hydrated. Figure 5 shows the collagen scaffold in the petri dish with and without the treatment of glutaraldehyde had three distinct exothermal peaks in the thermogram, with Tg is at 42 °C and 52 °C respectively. The total enthalpy for both conditions with and without glutaraldehyde crosslink is 97.14 $J.g^{-1}$ and 152.59 $J.g^{-1}$ respectively. The sample was thick and was not easily break. The collagen concentration is evenly distributed throughout the sample surface. The characteristics of scaffold in small aluminium tray that was treated with 1.0% GTA were preferred because it has the highest glass transition temperature compared to the other two scaffolds from large aluminium tray and petri dish. Tg will generally be low when the chain is highly flexible and it is higher when the chain is rigid. The results show that protein chain in scaffold from small aluminium tray that was treated with 1.0% GTA was rigid and this means that the scaffold was strong.

9 Conclusion

From the fabrication process of collagen scaffold, it can be concluded that the aluminium tray with dimension of 8.5 × 8.5 × 4 cm3 was appropriate to use with 33 ml of collagen solution with concentration of 5 mg/ml. Dehydrothermal treatment and 1.0% glutaraldehyde crosslinking greatly improved on the scaffold physical strength. The physical properties of the collagen scaffold fabricated using the above-mentioned condition was thick, strong and the collagen concentration was evenly distributed throughout the scaffold. The heat capacity was affected by thermal conductivity of the aluminium tray. The smaller the aluminium tray, the more effective it was to distribute heat. DSC analysis has shown that the glass transition temperature, Tg for small aluminium tray with DHT and 1.0% glutaraldehyde crosslinking was 145 °C. At temperature 145 °C, the collagen changes phase from glassy to rubbery or flexible thermoplastic state. It is noted that there appears to be close connection between the Tg value and the flexibility of the collagen chain. Since the Tg temperature was high then the collagen scaffold was rigid. From the graph, the shift peak for scaffold with glutaraldehyde crosslink was higher in both small aluminium tray and petri dish but it was not applicable for large aluminium tray.

This method showed that optimisation of conditions and physicochemistry of collagen scaffold can be controlled by using specific dimension of container and crosslinker. There are other several methods that can be applied in the fabrication of collagen scaffold to meet specific physicochemistry for specific biomaterial needs and functions. In this chapter, we would like to show one of the method that can be used in the fabrication of collagen scaffold for biomaterials. The ability of a country to produce technological advanced tissue engineering and regenerative medicine is vital for the country development to be sustainable and independent. Sustainable animal husbandry and agricultural practice for the production of optimised and functional collagen scaffold for tissue engineering and regenerative medicine applications are vital to meet the needs and demands for regenerative medicines which often are very costly and imported from outside ASEAN countries. It is important for ASEAN countries to reduce dependency from imported materials and work in collaboration with other ASEAN member countries. The production of regenerative scaffold and materials will increase the clinicians and patients' access to effective treatment and reduce the rate of mortality and morbidity associated with infections as well as reducing the rate of amputation.

Biotechnology from animal husbandry will benefit Malaysia and other ASEAN countries tremendously and encourage economic growth in other sectors such as agriculture and animal husbandry where the collagen is derived, biotechnology and pharmaceutical industries. We strongly believe that improvements and production of this optimised collagen scaffold will contribute to the greater benefit of Malaysian's health and economy's GDP.

With ASEAN Economic Cooperation (AEC), it is projected that Asia will be the second largest market for biomedical devices and healthcare after United States. Countries such as Thailand, Singapore, Malaysia and Indonesia are considered as hotspots for sales, investment and business opportunities. Therefore it is important for ASEAN to be able to produce their products to meet the demand of emerging market and create economic sustainability in the region.

The source of collagen from animals is also a serious concern for many population in ASEAN, where 60% of the population, approximately of 300 million people are Muslims and Hindus. Muslim and Hindu's cultural and religion sensitivities towards the products derived from porcine (pig) and bovine (cow) need to be addressed in order to penetrate the market. The invention and development of regenerative template for organ regeneration which have been successful for skin and nerve regeneration has improved lives of many people around the world, especially those of burnt and accident victims, offered new lease of life and improvement of life quality and health. It is important to note that the general public health sector is of importance in the development of a country and contributing directly to the national growth and Gross Domestic Produce (GDP).

The ability of a country to produce technological advanced tissue engineering and regenerative medicine is vital for the country development to be sustainable and independent. Through the development of halal derived collagen type I from ovine through sustainable animal husbandry and agricultural practice for the production of optimised and functional collagen scaffold for tissue engineering and regenerative medicine applications, ASEAN countries able to reduce dependency from imported materials and work in collaboration with other ASEAN member countries. The production of regenerative scaffold and materials will increase the clinicians and patients' access to effective treatment and reduce the rate of mortality and morbidity associated with infections as well as reducing the rate of amputation.

This tremendously encourage economic growth in other sectors such as agriculture and biopharmaceutical industries. We strongly believe that improvements and production of this optimised collagen scaffold will contribute to the greater benefit of Malaysian's health and economy's GDP.

References

Agochukwu, N. B., Wong, L., & Liau, J. Y. (2018). Dermal regenerative template as a cost-effective alternative for complex scalp reconstruction. *Journal of Craniofacial Surgery*.https://doi.org/10.1097/SCS.0000000000004068

Dong, C., & Lv, Y. (2016). Application of collagen scaffold in tissue engineering: Recent advances and new perspectives. In Polymers.https://doi.org/10.3390/polym8020042

Eltom, A., Zhong, G., & Muhammad, A. (2019). Scaffold techniques and designs in tissue engineering functions and purposes: A review. In *Advances in Materials Science and Engineering*.https://doi.org/10.1155/2019/3429527

FAOSTAT. (2016). Food agriculture organization database. FAOSTAT Domains, http://www.fao.org/faostat/en/#data/QL/visualize.

Harley, B. A. C., & Yannas, I. V. (2013). In vivo synthesis of tissues and organs. In *Principles of Tissue Engineering: Fourth Edition*. https://doi.org/10.1016/B978-0-12-398358-9.00018-5

Laurent, G. J. (1987). Dynamic state of collagen: Pathways of collagen degradation in vivo and their possible role in regulation of collagen mass. In *American Journal of Physiology—Cell Physiology*. https://doi.org/10.1152/ajpcell.1987.252.1.c1

Loh, T. C. (2004). Livestock production and the feed industry in Malaysia. *FAO Animal Production and Health Proceedings (FAO).*

San Juan, M. G., Bogdanski, A., & Dubois, O. (2019). Towards sustainable bioeconomy. *Lessons learned from case studies.*

Natural Bioactive Compounds in Safflower (*Carthamus tinctorius* L.) Seed, Seed Oil and Flower and Usages

Mohamed Elwathig Saeed Mirghani and Elfadl Yousif Elmoghtaba

Abstract Safflower (*Carthamus tinctorius* L.) is a plant that has been used for centuries for various purposes. It is primarily cultivated for its seeds, which are rich in natural bioactive compounds. This book section shall explore the bioactive compounds found in safflower seed, seed oil, and flower and their potential usage, which can maximize resource efficiency, minimize waste, promote sustainable cultivation practices, and create environmentally friendly products. *C. tinctorius* is a versatile plant known for its valuable bioactive compounds found in its seeds, seed oil, and flowers. The seeds are a rich source of omega-6 fatty acids, antioxidants, proteins, and fibre, offering benefits for skin health, brain function, and immune support. With its high linoleic acid and vitamin E content, Safflower seed oil finds applications in skin care products and culinary uses. The vibrant safflower flowers contain carthamin, a natural red pigment used for dyeing textiles and food colouring, while also potentially providing antioxidant and anti-inflammatory properties. Safflower and its derivatives are used in culinary applications, skincare products, natural dyeing, and traditional medicine, although further research is needed to fully explore their therapeutic potential to maximize the benefit by combining the principles of the circular economy with the utilization of biological resources.

Keywords Bioactive compounds · *Carthamus tinctorius* · Carthamin · Circular economy · And safflower

1 Introduction

Safflower (*Carthamus tinctorius* L.) is a plant that belongs to the family Compositae. Examples of historical and popular names of *Carthamus tinctorius* in various languages are safflower, dyer's thistle, false saffron, bastard saffron, and Dyer's saffron in English. It is also named Carthamus in Latin, osfur, usfar; qurtum, qorton;

M. E. S. Mirghani (✉) · E. Y. Elmoghtaba
National Oilseed Processing Research Institute (NOPRI), University of Gezira, P. O. Box 20, Medani, Gezira State, Sudan
e-mail: elwathig@gmail.com

F. Adyani Ahmad Fuad et al. (eds.), *Circular Bioeconomy*,
https://doi.org/10.1007/978-981-97-7010-6_4

khiri in Arabic; qurtami, qurtema, qurtam in Hebrew (Yue et al. 2013). It is a multipurpose oilseed crop grown mainly for its high quantity and quality edible oil, food additive, bird-feeding seed and other uses as a source of oil for the paint industry. Safflower is also, grown for its flowers, which are used for colouring as a natural pigment and flavouring foods, making dyes for the textile industry, livestock forage, vegetables, making herbal teas as well as for medicinal purposes.

The eastern part of the Mediterranean region is regarded as the original centre of this genus. The genus includes about 25 species, distributed from Spain and North Africa across the Middle East to northern India (Ashri & Knowles 1960). Safflower is also, grown as a medicinal plant and herbal medicine that is used for the treatment of cardiovascular diseases, male and female sterility, lowering blood cholesterol, the release of retained placenta and various types of rheumatism, respiratory diseases, gastritis, etc.

The composition of natural bioactive components in cultivated Safflower (*Carthamus tinctorius* L.) seed oils was studied by many researchers. Highly abundant bioactive compounds with hydrophilic nature, a lignan glycoside (tracheloside), flavonoids and alkaloids in seeds of many species, were identified. Only a minor part of the hydrophilic compounds present in the seeds was transferred into the seed oil during the extraction (Chakradhari et al. 2020). In the circular bioeconomy, these natural compounds can be efficiently extracted and utilized for various applications such as pharmaceuticals, nutraceuticals, cosmetics, and functional foods (Awasthi et al. 2022).

The α-Tocopherol was the main form of tocochromanols in safflower seed oils. βSitosterol was the predominant form of phytosterols, while high levels were also recorded for gramisterol and avenasterol in the seed oils. The total amount of minor lipophilic compounds such as tocochromanols, carotenoids and sterols in the seed oil was reported. There is some similarity in the fatty acid composition and levels in many varieties of safflower seed oil. The linoleic (>75%), oleic (>15%), palmitic (~5%), stearic (~2%) acids and others (>1%).

Despite the many uses of safflower, it has remained a minor crop. Therefore, the scientific community needs to carry out research on this crop and popularize it as a commercial crop for edible oil, development of pharmaceuticals, paint and varnishes industry, dye extraction (carthamin), source of α-tocopherol, livestock feed, vegetable and cut flower.

This chapter reviews the safflower bioactive compounds in the seed, cold-pressed safflower oil, and cake and their health effects, aroma compounds found in cold-pressed safflower oil, application areas, adulteration, and some process contaminants such as polycyclic aromatic hydrocarbons (PAHs) and 3-monochloropropanediol (3-MCPD) esters. This article also reviews the characteristic components that have been isolated from the florets of *C. tinctorius* such as quinochalcone, and *C*-glycosides.

The circular bioeconomy has the potential to contribute to a more sustainable and resilient economy by reducing resource depletion, greenhouse gas emissions, and pollution. It can also create new economic opportunities, generate green jobs, and support the transition to a low-carbon society.

2 The Plant

Safflower (*Carthamus tinctorius* L.) is a plant that belongs to the family Compositae. *Carthamus* L. is a genus belonging to the tribe *Cynareae* (thistle), subfamily *Tubuliflorae*. The eastern part of the Mediterranean region is the original centre of this genus. The genus includes about 25 species, distributed from Spain and North Africa across the Middle East to northern India (Ashri & Knowles 1960).

The names of the plant according to Aramaic/Hebrew are Qurtami, Qurtema, Qurtam Hebrew; Dardara & Qotzah (thorn, thistle); Moriqa (of the "thistle"). In Arabic, names are Osfur, Usfar; Qurtum, Qorton; Khiri. However, names in English are Safflower, Dyer's thistle, False saffron, Bastard saffron, and Dyer's saffron. (Yue et al. 2013). The chemical groups isolated from C. tinctorius included oils, proteins, minerals, phenolics, flavonoids, alkaloids, lignans, carboxylic acids, steroids, polysaccharides, quinochalcone C-glycosides and quinone-containing chalcones (Al-Snafi 2015a 2015b).

3 Seeds

The safflower (*C. tinctorius* L.) seed oil content ranges from 20 to 45%; the oil is high in linoleic acid, an unsaturated fatty acid that aids in lowering the blood cholesterol level. Thus, safflower has long been used as a medical plant in many countries, especially in China and India. However, for industrial purposes, it has long been neglected because of the low seed yield or oil content, until its physical role was revealed. In recent years, research works carried out in many countries mostly focused on improving the seed or oil yield (Liu et al. 2016).

Safflower seed is suitable for growing all over the world, with high soil and climate adaptation. The safflower seed is an important alternative oil crop because of its high oil content (27–32%) using cold press extraction (Ergönül & Özbek 2020). Oil content varied insignificantly with the study years, locations, sowing date, harvesting date and variety. It is also found that the oil content increased very rapidly from the 15th to the 30th day after the flowering sampling period and reached its maximum level of maturity in both varieties… Generally, and at maturity, oil content for many varieties in the same location was higher in spring sowing than that observed in autumn sowing (Gecgel et al. 2007). However, numerous research studies have shown that oil content increases during safflower seed development, and climatic factors during the growing season, maturity, variety, and location influence oil concentration (Esendal & Tosun 1972; Hamrouni et al. 2004; Nagaraj & Reddy 1997; Paşa 2008).

4 Oil

The safflower plant (*Carthamus tinctorius* L.) is suitable for growing all over the world, with high soil and climate adaptation. This plant is cultivated mainly for its seed, which is used as edible oil. The most common volatile compounds found in cold-pressed safflower seed oil are phenol-asarone, β-asarone, 1-methyl-4-(2-propenyl)-benzene, diethyl-benzene, acetoin, methyl benzene, hexanal, p-xylene, heptanal, and 2-octenal (Ergönül & Özbek 2020).

Safflower seed oil is rich in linoleic acid (55–70%) and unsaturated fatty acids that benefit in various uses such as medicinal and/or industrial. Cold-pressed safflower oil possesses high nutritional and pharmaceutical values due to its noticeable amounts of bioactive compounds and essential fatty acids. The oil is a valuable source of α-tocopherol, which shows the highest vitamin E activity, and it, therefore, has many health benefits such as prevention and treatment of hyperlipemia, arteriosclerosis, and coronary heart disease (Kochhar 2011).

Al Surmi et al. (2015) reported that safflower seeds contain a high content of edible oil which is considered a good source of essential linoleic fatty acid. The fatty acid composition of the oils indicated that the predominant fatty acid was linoleic acid (74.60–78.24%) followed by oleic (11.22–14.19%), palmitic (6.03–6.66%) and stearic (2.01–2.61%).

However, salinity stress caused affected the fatty acid composition by a significant increase in oleic acid (C18:1) and significant decreases in linoleic (C18:2) and linolenic (C18:3) acids (Yeilaghi et al. 2012). The total lipids showed the per cent of triglyceride (81.70–85.34%), total hydrocarbons constituted from 88.0 to 94.7% while total sterols ranged from 5.32 to 12.00% of the total unsaponifiable matters of the oils. The tocopherol content ranged from 1.36 to 56.96 mg/100 g oil. Fatty acid Composition of the oil from the seeds of different varieties of safflower (*C. tinctorius* L.) was also reported by Gecgel et al. (2007) in a study that showed the important factors influencing fatty acid composition are seed genotype and sowing date. So, safflower optimum sowing date should be based on a detailed understanding of the environmental factors influencing yield components, oil quality, fatty acid composition, and of course also its genotype-dependent characteristics.

5 Bioactive Compounds

Safflower (*Carthamus tinctorius* L.) seeds were evaluated for melanogenesis inhibitory activity and its active principles were identified following activity-guided isolation to develop a new skin-whitening agent (Roh et al. 2004). Serotonin derivatives were extracted from the fractionation of the MeOH extract of *C. tinctorius* oil cake (Nagatsu et al. 1998).

Cai et al. (2004) reported that *Carthamus tinctorius* L. ethanolic extract has Trolox Equivalent Antioxidant Capacity (TEAC) 318.0 and 207.4 μmol trolox/100 gDW, for methanolic and water extracts, respectively. The same authors (Cai et al. 2004) also reported the total phenolic contents were 2.12 and 1.32 g/100 gDW for methanolic and water extracts, respectively. The major types, (representative compounds), for total phenolics, were flavonoids (chalcones: carthamin, carthamone), and lignin.

Yue et al. (2013) Isolated Quinochalcone C-glycosides as characteristic components from the florets of *Carthamus tinctorius.* Quinochalcone C-glycosides were found to have multiple pharmacological activities, which has attracted the attention of many researchers to explore these compounds.

Moreover, a report focusing on the biological activities of safflower natural quinochalcones is influenced by environmental factors. These findings underlined the potential consumption of safflower as a suitable source of natural dyes as an alternative to food synthetic colourants (Salem et al. 2014). To summarize quinochalcone C-glycosides' physicochemical properties, chromatographic behaviour, spectroscopic characteristics, as well as their biological activities, which will be helpful for further study and development of quinochalcone C-glycosides. By maximizing the extraction and utilization of these compounds, we can minimize waste and efficiently optimize resources, utilizing biomass and environmentally friendly sustainable production as the main principles of circular bioeconomy (Muscat et al. 2021).

6 Antioxidants

There are many shreds of evidence that natural products and their derivatives have efficient antioxidative characteristics, consequently linked to anticancer, hypolipidemic, antiaging and anti-inflammatory activities (Aruoma 2003; Wiseman et al., 1997; Cho et al. 2006; Hogg 1998; Mirghani et al. 2018; Kruawan & Kangsadalampai 2006).

As reported by Asgrapanah and Kazemivash (2013), the antioxidative capacities of *C. tinctorius* were evaluated by determining its effect on 2,2-diphenyl-1- picrylhydrazyl (DPPH) radical scavenging and ferric reduction. In both assays, the aqueous extract of the flowers exhibited high antioxidant activity. DPPH scavenging effect was 96.65%. The 50% inhibitory concentration (IC_{50}) value the for Ferric reduction assay was determined as 1,140.5 μmol/g (Kruawan & Kangsadalampai 2006). The total phenolic content from the flowers was determined as 2.12 and 1.32 g/100 g for methanolic and aqueous extracts respectively (Cai et al. 2004). As the flowers contained high phenolic compounds including it confirmed that they have an important role in antioxidant activities as reported by Calomme et al. (1996). The oxidative stability of the studied oils ranged from 6.20 to 6.45 h by using Rancimat at 100 °C as reported by Al Surmi et al., (2015). Hiramatsu et al. (2009) studied the antioxidant and Neuroprotective Activities of Safflower, (*Carthamus tinctorius* Linne). Their study found that the petal extract of *C. tinctorius* has free radical scavenging activity and neuroprotective effect and the major active components are carthamin and polyphenols. They concluded that petals might be an excellent antioxidant ingredient to protect against neuronal damage induced by free radicals involved in ageing associated with neuron loss, dementia including Alzheimer's disease, stroke and other neurogenerative diseases.

7 Uses

Saito and Mori (1994), used the powder florets from Dyer's saffron (*Carthamus tinctorius* L.) by suspending them in a buffered solution containing $KMnO_4$ or H_2O_2 that oscillated with a sonicator at ambient temperature. Carthamin yield in the floral pastes was determined showing that at short-time sonication, carthamin yield could be increased by about 8% higher on average.

For a long time, *C. tinctorius* has been used in traditional medicines as a purgative, analgesic, antipyretic and antidote to poisoning (Asgarpanah & Kazemivash 2013). It also has been shown to have antioxidant, analgesic, anti-inflammatory and antidiabetic activities. the oil is high in linoleic acid, an unsaturated fatty acid that aids in lowering the blood cholesterol level. Thus, safflower has long been used as a medical plant in many countries, especially in China and India (Liu et al. 2016).

Natural food pigments such as carotenoids, anthocyanins, flavonoids, anthraquinone and others are utilized to colour food, it is a requirement to indicate what is used if the pigment remains in the final products in a detectable amount. Many analytical techniques could be used to identify and quantify the amount of food pigment, however, Watanabe and Terabe (2000) used capillary electrophoresis (CE) that shown to be a useful technique for the analysis of these natural food pigments and the pigments extracted from commercial food samples by solidphase extraction method. In Egypt, dye from safflower was used to colour cotton and silk as well as ceremonial ointment used in religious ceremonies and to anoint mummies before binding. Egyptian safflower dye was used in Italy, France and Britain to colour cheese and flavour sausage (Al Surmi et al. 2015). Carthamin was dissolved in alkaline solutions and the colour stability was examined by the spectrophotometric-assay technique. The experimental results suggest that the red colour of carthamin can be retained for 24 h or so when treated pertinently under alkaline conditions (Saito & Mori 1994).

Sun et al. (2014) optimized the extraction conditions of natural colourant carthamin from safflower (*C. tinctorius* L.) by response surface methodology using alkali-solution and acid-isolation method (ASAI) and aqueous two-phase system separation method (ATSS). The two methods were compared based on the optimisation results of extraction conditions by response surface methodology (RSM). In ASAI, the maximum extraction yield reached 0.779% at a pH value of potassium carbonate solution of 11.16, the ratio of potassium carbonate solution to the raw material of fifteen and extraction time of 18 min, while 2.652% was achieved with a concentration of acetone of 58%, ratio of acetone solution to the raw material of twenty-three and extraction time of 41 min in ATSS. From the point of view of extraction yield, ATSS had more superiority to ASAI.

Bae et al. (2002) reported that safflower seed oil has a high linoleic acid level, which possesses anti-inflammatory activity in bone by moderating prostanoid formation, correcting bone loss due to ovariectomy and increasing intestinal calcium absorption having a good effect on osteoporosis. The same authors (Bae et al. 2002) also reported that safflower seed powder contains many minerals, especially calcium, magnesium and potassium, and is effective in preventing the osteoporotic process caused by bilateral ovariectomy in a rat model. It was verified that the seed powder effectively inhibited bone loss associated with estrogenic deficiency in rats.

The flower aqueous extract can reverse the metabolic disorders occurring in alloxan-induced diabetes. Considering these effects on these lipid components, it can be assumed as a potential hypolipidemic agent, which will be a great advantage both in diabetic conditions as well as associated atherosclerosis or hyperlipidemic conditions (Asgary et al. 2012). *C. tinctorius* flowers regenerate and restore Langerhans islets, thus the insulin level would be elevated. Safflower enhances the secretion of insulin from the beta cells of the islets of Langerhans which makes it useful for diabetes and its complications. Safflower oil and folic acid supplementations have been shown to reduce embryo malformations in experimental models of diabetes as reported by Higa et al. (2012), folic acid and safflower oil supplementations interact and protect the embryos from diabetes-induced damage through several pathways related to decrease in proinflammatory mediators.

8 Flower

C. tinctorius is an important source of the tinctorially useful dyestuff, carthamin, which is accumulated in the tubular flowers through an enzyme process (Calomme et al. 1996; Saito et al. 1983). Carthamin and safflower yellow are the main constituents in the flower of *C. tinctorius.* Carthamidin, isocarthamidin, hydroxysafflor yellow A, safflor yellow A, safflamin C and luteolin are the main constituents which are reported from this plant (Asgarpanah & Kazemivash 2013). The same authors also reported that caryophyllene, p-allyl toluene, 1-acetoxytetralin and heneicosane were identified as the major components of *C. tinctorius* flowers' essential oil.

Salem et al. (2014) investigated the natural dyes contents of *C. tinctorius* and its biological activities. Obtained results showed that quinochalcone contents and antioxidant activities varied considerably as a function of flowering stages. So, flowers at the fructification stage contained the highest carthamin content with the strongest antioxidant capacity with all assays (FRAP, DPPH, and chelating power methods). The quantitative variation of these molecules could be due to the colour change of *C. tinctorius* flowers. Correlation analysis indicated that the ABTS method showed the highest correlation coefficients with carthamin and precarthamin contents, that is, 0.886 and 0.973, respectively. During flowering, the antimicrobial activity of these two natural dyes increased where the maximum inhibitory effect mentioned with carthamin mainly against *E. coli* (iz = 25.89 mm) at the fructification stage. Therefore, the increased frequency of resistance to commonly used antibiotics leads to the search for new effective natural drugs in the food and pharmaceutical industries.

Water-soluble yellow and water-insoluble red carthamin pigments can be extracted from safflower petals, having the potential as a food colourant' and natural red dye. In the future natural valuable food dyes will continue to be widely accepted in food products due to their non-allergic and non-carcinogenic properties with better stability over pH, temperature and different solute concentrations and the presence of polar solvents (Jadhav & Joshi 2015).

9 Carthamin

Carthamin is a natural red pigment derived from safflower and has been utilized in industries such as dye, food colouring, and natural medicine. Carthamin is a well-known and useful red dyestuff, recently attracting attention as one of many harmless bio-colourants for processed foods. Reproducing the complex biosynthesis of carthamin has been a common research goal. Carthamin has been rendered complex because of the unstable precursors and low yields from the multiple steps involved in the synthesis. The synthesis involves mild conditions, short reaction time, and readily available compounds. After de-*O*-methylation, condensation, oxidative decarboxylation, and saponification, carthamin was obtained with an overall yield of 9.5% through the proposed method (Abe et al. 2020).

Li et al. (2009) investigated the effects of carthamins yellow (CY) on a blood stasis model, which was obtained by placing rats in ice-cold water during the time interval between two injections of epinephrine. The results demonstrated that the CY significantly decreased the whole blood viscosity, plasma viscosity, and erythrocyte aggregation index, which were increased in the blood stasis model. Therefore, CY administration might provide the additional benefit of increasing blood fluidity by lowering blood viscosity, which can be of great value in the prevention of hemorheological disorder-associated diseases in at-risk patients.

In vitro cultured cells of the flowering plant give no carthamin; however, they retain a potential activity to produce another red pigment, kurenamin, at a considerably high level (Saito et al. 1988). This phenotypic difference seems to be helpful for us to find a clear-cut distinction in the metabolic diversity between intact tissues and cultured cells. Only through comparative studies, was some notable evidence obtained, showing that the imaginary biochemical modification is practically introduced into the cell cultures during the special cultivation.

A previous study (Shimokoriyama & Hattori 1955) on the process of reddening of yellow flower petals of *Carthamus tinctorius* was biochemically investigated and it was concluded that the process may involve the polyphenol oxidase- or peroxidase-catalyzed oxidation of a leucopigment, i.e. dihydrocarthamin, as a precursor.

10 Conclusion

Due to the easy collection of the safflower (*C. tinctorius*) plant and being widespread and also remarkable biological activities, this plant has become both food and medicine in many parts of the world. *C. tinctorius* is a plant with a wide range of chemical constituents which exerted many pharmacological effects. There is a great promise to develop novel drugs from petals and other parts of the plant to treat human diseases due to their effectiveness and safety.

The information provided in this review shows that *C. tinctorius* has pharmacological functions including antioxidant, anti-inflammation, analgesic, antidiabetic, hepatoprotective and antihyperlipidemic activities. However, many of the present experimental research work and reports suggested a possibility that serotonin derivatives and flavonoids can be further developed as a potential disease therapeutic medicine. As a result, a further stream of information shows, it is also possible that many compounds extracted from *C. tinctorius* such as furanocoumarins, might be useful in the development of new drugs to treat various diseases.

The findings also, revealed that *C. tinctorius* flowers at the fructification stage showed the highest carthamin content with the strongest antioxidant and significant antimicrobial activities, suggesting that this stage was the best harvesting time for safflower. These findings underlined the potential consumption of safflower as a suitable source of natural dyes as an alternative to food synthetic colourants.

For these reasons, extensive pharmacological and chemical experiments, together with human metabolism will be a focus for future research studies and applications on the potential of *C. tinctorius* to be employed in new therapeutic drugs starting from the farm by applying the principals of circular bioeconomy has the potential to contribute to a more sustainable and resilient economy by reducing resource depletion, greenhouse gas emissions, and pollution. It can also create new economic opportunities by promoting the efficient use of biological resources, generate green jobs, reducing waste and support the transition to a low-carbon society, minimizing environmental impact.

References

Abe, Y., Sohtome, T., & Sato, S. (2020). Biomimetic synthesis of carthamin, a red pigment in safflower petals, via oxidative decarboxylation. *Journal of Heterocyclic Chemistry, 57*(10), 3685–3690. https://doi.org/10.1002/jhet.4089

Al-Snafi, A. E. (2015a). Chemical constituents and pharmacological importance of Agropyron repens–a review. *Research Journal of Pharmacology and Toxicology, 1*(2), 37–41. http://www.asdpub.com/index.php/rjpt

Al-Snafi, A. E. (2015b). The chemical constituents and pharmacological importance of *Carthamus tinctorius*-an overview. *Journal of Pharmaceutical Biology, 5*(3), 143–166. https://www.researchgate.net/profile/Ali-Al-Snafi/publication/313678942

Al Surmi, N. Y., El Dengawi, R. A. H., Khalefa, A. H., & Yahia, N. (2015). Characteristics and oxidative stability of some safflower (*Carthamus Tinctorius* L.). *Journal of Nutrition & Food Sciences, 14*, 2. https://doi.org/10.4172/2155-9600.S14-001

Aruoma, O. I. (2003). Methodological considerations for characterizing potential antioxidant actions of bioactive components in plant foods. *Mutation Research/fundamental and Molecular Mechanisms of Mutagenesis, 523*, 9–20. https://doi.org/10.1016/S0027-5107(02)00317-2

Asgarpanah, J., & Kazemivash, N. (2013). Phytochemistry, pharmacology and medicinal properties of *Carthamus tinctorius* L. *Chinese Journal of Integrative Medicine, 19*(2), 153–159. https://doi.org/10.1007/s11655-013-1354-5

Asgary, S., Rahimi, P., Mahzouni, P., & Madani, H. (2012). Antidiabetic effect of hydroalcoholic extract of *Carthamus tinctorius* L. in alloxan-induced diabetic rats. *Journal of Research in Medical Sciences: the official journal of Isfahan University of Medical Sciences, 17*(4), 386. https://www.ncbi.nlm.nih.gov/pmc/articles/PMC3526135/pdf/JRMS-17-386.pdf

Ashri, A., & Knowles, P. F. (1960). Cytogenetics of safflower (*Carthamus* L.) species and their hybrids 1. *Agronomy Journal, 52*(1), 11–17. https://doi.org/10.2134/agronj1960.00021962005200010004x

Awasthi, M. K., Singh, E., Binod, P., Sindhu, R., Sarsaiya, S., Kumar, A., & Zhang, Z. (2022). Biotechnological strategies for bio-transforming biosolid into resources toward circular bio-economy: A review. *Renewable and Sustainable Energy Reviews, 156*, 111987. https://doi.org/10.1016/j.rser.2021.111987

Bae, C. S., Park, C. H., Cho, H. J., Han, H. J., Kang, S. S., Choi, S. H., & Uhm, C. S. (2002). Therapeutic effects of safflower (*Carthamus tinctorius* L.) seed powder on osteoporosis. *Applied Microscopy, 32*(3), 285–290. https://www.researchgate.net/profile/Seok-Hwa-Choi/publication/264141146

Cai, Y., Luo, Q., Sun, M., & Corke, H. (2004). Antioxidant activity and phenolic compounds of 112 traditional Chinese medicinal plants associated with anticancer. *Life Sciences, 74*(17), 2157–2184. https://doi.org/10.1016/j.lfs.2003.09.047

Calomme, M., Pieters, L., Vlietinck, A., & Berghe, D. V. (1996). Inhibition of bacterial mutagenesis by Citrus flavonoids. *Planta Medica, 62*(03), 222–226. https://doi.org/10.1055/s-2006-957864

Chakradhari, S., Perkons, I., Mišina, I., Sipeniece, E., Radziejewska-Kubzdela, E., Grygier, A., & Górnaś, P. (2020). Profiling of safflower seeds and seed oil bioactive components: Cultivated vs. wild. *European Food Research and Technology, 246*, 449–459. https://doi.org/10.1007/s00217-019-03414-w

Cho, J. Y., Park, S. C., Kim, T. W., Kim, K. S., Song, J. C., Lee, H. M., Lee, C. H., et al. (2006). Radical scavenging and anti-inflammatory activity of extracts from Opuntia humifusa Raf. *Journal of Pharmacy and Pharmacology, 58*(1), 113–119.https://doi.org/10.1211/jpp.58.1.0014

Ergönül, P. G., & Özbek, Z. A. (2020). Cold pressed safflower (*Carthamus tinctorius* L.) seed oil. *Cold Pressed Oils*, 323–333. https://doi.org/10.1016/B978-0-12-818188-1.00029-3

Esendal, E., & Tosun, F. (1972). Research on the yield, phonological, morphological and seed characteristics of some native and foreign varieties of safflower (*Carthamus tinctorius* L.) under Erzurum ecological conditions. *Ataturk Univ J Agric Fac, 3*, 93–115. https://www.cabidigitallibrary.org/doi/full/10.5555/19740720819

Gecgel, U., Demirci, M., Esendal, E., & Tasan, M. (2007). Fatty acid composition of the oil from developing seeds of different varieties of safflower (*Carthamus tinctorius* L.). *Journal of the American Oil Chemists' Society, 84*(1), 47–54. https://doi.org/10.1007/s11746-006-1007-3

Hamrouni, I., Touati, S., Dhifi, W., Chahed, T., Ayachi, S., Salah, H. B., & Marzouk, B. (2004). Glycerolipid evolution during safflower seed formation and ripening. *Journal of Food Lipids, 11*(4), 297–311. https://doi.org/10.1111/j.1745-4522.2004.11410.x

Higa, R., Kurtz, M., Mazzucco, M. B., Musikant, D., White, V., & Jawerbaum, A. (2012). Folic acid and safflower oil supplementation interacts and protects embryos from maternal diabetes-induced damage. *Molecular Human Reproduction, 18*(5), 253–264. https://doi.org/10.1093/molehr/gar080

Hiramatsu, M., Takahashi, T., Komatsu, M., Kido, T., & Kasahara, Y. (2009). Antioxidant and neuroprotective activities of Mogami-benibana (safflower, *Carthamus tinctorius* Linne). *Neurochemical Research, 34*(4), 795–805. https://doi.org/10.1007/s11064-008-9884-5

Hogg, N. (1998). *Free radicals in disease.* Paper presented at the Seminars in reproductive endocrinology.https://doi.org/10.1055/s-2007-1016284

Jadhav, B. A., & Joshi, A. A. (2015). Extraction and quantitative estimation of bioactive component (yellow and red carthamin) from dried safflower petals. *Indian Journal of Science and Technology, 8*(16), 1–5. https://citeseerx.ist.psu.edu/document?repid=rep1&type=pdf&doi=97ad9a7a43ae8db817cf66ba8a146ceaf74a00cb

Kochhar, S. P. (2011). 11 Minor and speciality oils. In *Vegetable oils in food technology: Composition, properties and uses* (p. 291). Wiley. ISBN: 1444339915, 9781444339918.

Kruawan, K., & Kangsadalampai, K. (2006). Antioxidant activity, phenolic compound contents and antimutagenic activity of some water extract of herbs. *The Thai Journal of Pharmaceutical Sciences, 30*(1), 2835. https://digital.car.chula.ac.th/tjps/vol30/iss1/4

Li, H. X., Han, S. Y., Wang, X. W., Ma, X., Zhang, K., Wang, L., & Tu, P. F. (2009). Effect of the carthamins yellow from *Carthamus tinctorius* L. on hemorheological disorders of blood stasis in rats. *Food and chemical toxicology, 47*(8), 1797–1802. https://doi.org/10.1016/j.fct.2009.04.026

Liu, L., Guan, L. L., & Yang, Y. X. (2016). A review of fatty acids and genetic characterization of safflower (*Carthamus tinctorius* L.) seed oil. *World Journal of Traditional Chinese Medicine, 2*(2), 48. https://doi.org/10.15806/j.issn.2311-8571.2016.0006

Mirghani, M. E., Elnour, A. A., Kabbashi, N. A., Alam, M. Z., Musa, K. H., & Abdullah, A. (2018). Determination of the antioxidant activity of gum Arabic: An exudation from two different locations. *Sci Asia, 44*(2018), 179–186. https://doi.org/10.2306/scienceasia1513-1874.2018.44.179

Muscat, A., de Olde, E. M., Ripoll-Bosch, R., Van Zanten, H. H., Metze, T. A., Termeer, C. J., & de Boer, I. J. (2021). Principles, drivers and opportunities of a circular bioeconomy. *Nature Food, 2*(8), 561–566. https://doi.org/10.1038/s43016-021-00340-7

Nagaraj, G., & Reddy, P. S. (1997, June). Some factors influencing safflower seed and oil quality. In *Proceedings of the International Safflower Conference. Bari, Italy* (pp. 347349). https://www.jstage.jst.go.jp/article/cpb1958/46/6/46_6_1044/_pdf

Nagatsu, A., Zhang, H. L., Watanabe, T., Taniguchi, N., Hatano, K., Mizukami, H., & Sakakibara, J. (1998). New steroid and matairesinol glycosides from safflower (*Carthamus tinctorius* L.) oil cake. *Chemical and Pharmaceutical Bulletin, 46*(6), 10441047. https://doi.org/10.1248/cpb.46.1044

Paşa, C. (2008). *Kışlık ve yazlık ekimin aspir (Carthamus tinctorius L.) bitkisinin verimi ve bitkisel özelliklerine etkisi* (Master's thesis, Namık Kemal Üniversitesi). https://acikbilim.yok.gov.tr/handle/20.500.12812/669178

Roh, J. S., Han, J. Y., Kim, J. H., & Hwang, J. K. (2004). Inhibitory effects of active compounds isolated from safflower seeds for melanogenesis. *Biological and Pharmaceutical Bulletin, 27*, 1976–1978. https://doi.org/10.1248/bpb.27.1976

Saito, K., Daimon, E., Kusaka, K., Wakayama, S., & Sekino, Y. (1988). Accumulation of a novel red pigment in cell suspension cultures of floral meristem tissues from *Carthamus tinctorius L. Zeitschrift für Naturforschung C, 43*(11–12), 862–870.https://doi.org/10.1515/znc-1988-11-1212

Saito, K., & Mori, T. (1994). Colour stability of carthamin under alkaline conditions. *Food Chemistry, 51*(1), 105–107. https://doi.org/10.1016/0308-8146(94)90056-6

Saito, K., Takahashi, Y., & Wada, M. (1983). Enzymic synthesis of carthamin in safflower. *Biochimica et Biophysica Acta-General Subjects, 756*(2), 217–222. https://doi.org/10.1016/0304-4165(83)90094-6

Salem, N., Msaada, K., Elkahoui, S., Mangano, G., Azaeiz, S., Ben Slimen, I., & Marzouk, B. (2014). Evaluation of antibacterial, antifungal, and antioxidant activities of safflower natural dyes during flowering. *BioMed Research International, 2014*. https://doi.org/10.1155/2014/762397

Shimokoriyama, M., & Hattori, S. (1955). On the formation of carthamin in the flowers of *Carthamus tinctorius*. *Archives of Biochemistry and Biophysics, 54*(1), 93–101. https://doi.org/10.1016/0003-9861(55)90011-4

Sun, Y., He, Y., Gu, M., Chen, J., & Li, F. (2014). Optimisation of the extraction conditions of natural colourant carthamin from safflower (*Carthamus tinctorius* L.) by response surface methodology. *International Journal of Food Science & Technology, 49*(4), 1168–1174. https://doi.org/10.1111/ijfs.12413

Watanabe, T., & Terabe, S. (2000). Analysis of natural food pigments by capillary electrophoresis. *Journal of Chromatography A, 880*, 311–322. https://doi.org/10.1016/S0021-9673(00)00209-0

Yeilaghi, H., Arzani, A., Ghaderian, M., Fotovat, R., Feizi, M., & Pourdad, S. S. (2012). Effect of salinity on seed oil content and fatty acid composition of safflower (*Carthamus tinctorius* L.) genotypes. *Food Chemistry, 130*(3), 618–625. https://doi.org/10.1016/j.foodchem.2011.07.085

Yue, S., Tang, Y., Li, S., & Duan, J. A. (2013). Chemical and biological properties of quinochalcone C-glycosides from the florets of *Carthamus tinctorius*. *Molecules, 18*, 15220–15254. https://doi.org/10.3390/molecules181215220

Energy

Characterization of Biodiesel Produced from Multiple Feedstocks for Sustainable Production of Agro-Based Fuels

Muhammad Amirul Ashraf Razali, Farah B. Ahmad, and H. H. Masjuki

Abstract In this study, biodiesel was derived from multiple vegetable oils mixed with palm oil. Transesterification of vegetable oil was performed using methanol and alkaline catalyst (potassium hydroxide (KOH)) at varying temperature (50–70 °C). The resulting biodiesel or fatty acid methyl esters (FAME) was evaluated for its chemical bonds and different FAME composition, in which the results from the latter were used to determine its fuel properties. The physico-chemical properties of biodiesel from multiple feedstocks were analyzed empirically for calorific value, cetane number, density, and kinematics viscosity. Biodiesel produced exhibited improved performance in some fuel properties in comparison to biodiesel prepared from single vegetable oil feedstock, especially for cetane number values and kinematic viscosity.

Keywords Biodiesel · Cetane number · Coconut oil · Corn oil · Palm oil · Transesterification

1 Introduction

Due to its reliability and efficiency, diesel engines play a significant role in the industrial energy and automotive sectors. However, diesel fuel emits high amount emissions that bring detrimental effect to the environment. An alternative fuel to be used in diesel engine is needed to reduce the dependency on fossil fuel. Biodiesel

M. A. A. Razali · H. H. Masjuki
Faculty of Engineering, Department of Mechanical Engineering, International Islamic University Malaysia, Kuala Lumpur, Malaysia

F. B. Ahmad (✉)
Faculty of Engineering, Department of Biotechnology Engineering, International Islamic University Malaysia, Kuala Lumpur, Malaysia
e-mail: farahahmad@iium.edu.my

F. B. Ahmad · H. H. Masjuki
Center for Energy Science, Department of Mechanical Engineering, University of Malaya, 50603 Kuala Lumpur, Malaysia

F. Adyani Ahmad Fuad et al. (eds.), *Circular Bioeconomy*,
https://doi.org/10.1007/978-981-97-7010-6_5

is promising as a new source of fuel that can be used in diesel engine, with more sustainable, greener production and lower emission of pollutants in comparison to diesel fuel (Martani et al. 2020). Biodiesel composed of long-chain alkyl esters formed by reacting fatty acids from oil or lipids with alcohol, typically via catalyzed transesterification process (Sudrajat et al. 2020). Transesterification is a method of converting a fatty acids of vegetable oil using alcohol (usually methanol or ethanol) that can be catalyzed by acids, bases, or enzymes to give alkyl esters (fatty acid methyl ester (FAME) or fatty acid ethyl ester (FAEE)) or biodiesel and glycerol as by-products (Fig. 1). Biodiesel can be used without modification in diesel engines. Vegetable oil feedstock is the most important of these materials as it plays a decisive role in biodiesel costs. However, there are limited studies on biodiesel resulting from transesterification of multiple feedstocks of vegetable oil.

Numerous vegetable oils feedstock were used the synthesis of biodiesel in various countries depending on its availability (Demirbas 2009a). Palm oil-derived biodiesel production is favorable as long as palm oil production is sustainable to also meet the food security demand. Oil palm is an oil crop that produces highest yielding per hectare on Earth, with superior oil yield per hectare in comparison to other well-known vegetable oil, for example sunflower and soybean (Ahmad et al. 2019). The application of biodiesel derived from palm oil in internal combustion engine requires only minimal engine modification (Samsukumar et al. 2015). Palm oil-derived diesel blend was reported not to compromise engine life and to impose less adverse effect to human health and the environment (Samsukumar et al. 2015). Corn oil is another vegetable oil that is highly available for the production of biodiesel. Corn is widely grown in the USA, in which the plantation of corn oil is less costly and more feasible

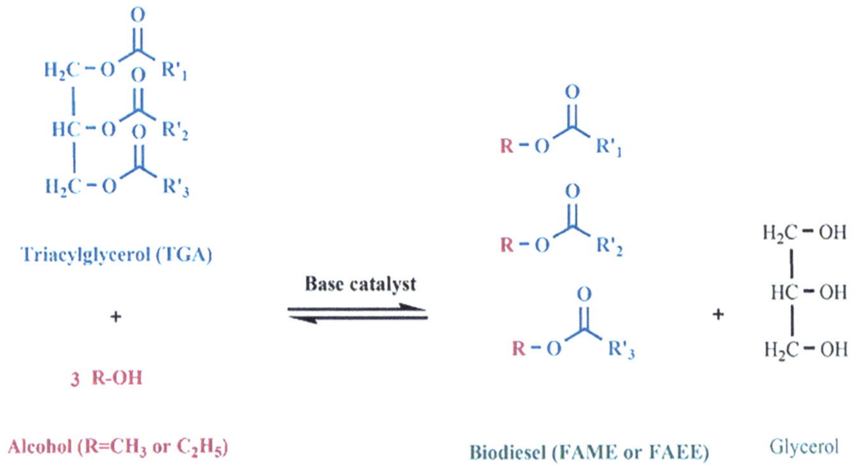

Fig. 1 Transesterification reaction of TGA, found from fatty acids of plant oils or lipids, and alcohol (e.g., methanol or ethanol) using homogenous base catalyst (e.g., KOH or NaOH) for the production of biodiesel (FAME of FAEE, respective to the alcohol reagent used) and glycerol as the by-product (Ahmad et al. 2018; Hazmi et al. 2021)

in comparison to other crops in the USA (Manigandan et al. 2019). The price of corn oil vegetable is comparable to untaxed petrol diesel (Manigandan et al. 2019). The fuel properties of corn biodiesel are comparatively similar to petrol diesel (Annamalai et al. 2018), especially on kinematic viscosity, density and cetane number. Coconut oil is another promising commercial feedstock for biodiesel production. Coconut palm is also another high oil yielding crop (Habibullah et al. 2014), in which 85–90% of its oil content composed of highly saturated and short chains fatty acids (Bongfa et al. 2020). Therefore, the use of coconut oil as feedstock for biodiesel synthesis may lead to fuel performance enhancement due to better ignition quality and oxidation stability. The application of coconut oil biodiesel lead to lowering air pollutants emissions, including black smoke, nitrous oxide and sulfur oxide (Cooperation 2009).

Biodiesel feedstock selection from vegetable oil is important as the right decision will ensure the success of commercial scale operation in biodiesel synthesis (Anwar et al. 2019), in terms of the availability of the vegetable oil in the global market and potential improvement of fuel properties. This study aims to characterize biodiesel produced from multiple feedstocks of palm oil, corn oil, and coconut oil for physical, chemical, and fuel properties.

2 Experimental Procedure

2.1 *Biodiesel Preparation*

A mixture of vegetable oil from coconut, corn, and palm oil was used for the conversion biodiesel via transesterification process. The selection of ratio for the oil mixture at 5:4:1 (corn oil: palm oil: coconut oil).

Transesterification process undertaken with methanol (20% w/w) was catalyzed by potassium hydroxide (KOH) (1% w/w) as base catalyst (Chiou et al. 2008), performed for 3 h with 500 rpm stirring and varying reaction temperatures (50–70 °C). FAME was separated from crude biodiesel with glycerol as by-product. The yields of biodiesel were calculated according to Eq. (1).

$$\text{Biodiesel yield} = \frac{\text{weight of FAME } (g)}{\text{weight of oil mixture } (g)} \times 100\% \tag{1}$$

2.2 *FAME Composition Analysis*

Gas chromatography-mass spectrometry (GC–MS) was performed to identify the composition of each methyl ester of fatty acid formed in biodiesel after the transesterification process. Biodiesel sample was placed in hexane solvent for GC–MS run. The column used was HP-5 (Agilent Technologies, California, United States), the carrier gas was helium with a flow rate at 1 ml/min. The split ratio was 2:10 and the injector temperature was fixed at 320 °C (Pontoh 2016). The temperature of the column was set at 150 °C and was held for 1 min (Pontoh 2016). The temperature was then increased at 15 °C/min to 320 °C and was held for 8 min (Pontoh 2016).

2.3 Chemical Bonds and Chemical Functional Groups Analysis

Fourier-Transform Infrared spectroscopy (FTIR) analysis was carried out using the FTIR spectrophotometer (NicoletTM iS50 FTIR, ThermoFisher Scientific, Massachusetts, United States). The spectral range was set at 4000–400 cm^{-1} and 4 cm^{-1} resolution with 20 scans (Zicker et al. 2016).

2.4 Analysis of Biodiesel Composition and Its Fuel Properties

Fuel properties including calorific value, cetane number, density, and kinematics viscosity of fatty acid methyl esters were calculated empirically based on the equations obtained by Ramírez-Verduzco et al. (2012) as presented in Eqs. (2), (3), (4) and (5) respectively (Ahmad et al. 2016; Ramírez-Verduzco et al. 2012). In these empirical formula, M represents the molecular weight of the fatty acid in FAME, and N is the number of double bonds in the respective fatty acid (Ramírez-Verduzco et al. 2012). Therefore, N could be equal to zero, one, two, three for the corresponding fatty acid C18:0, C18:1, C18:2, and C18:3.

The cetane number of biodiesel was calculated from

$$\emptyset = -\mathbf{7.8} + (\mathbf{0.302}).(\boldsymbol{M}) - (\mathbf{20}).(\boldsymbol{N}) \tag{2}$$

where Ø is the cetane number of FAME. The kinematics viscosity of biodiesel was obtained from

$$\varphi = e^{[-\mathbf{12.503}+(\mathbf{2.496}).\mathbf{ln}(\boldsymbol{M})-(\mathbf{0.718}).(\boldsymbol{N})]} \tag{3}$$

where φ is the kinematics viscosity at 40 °C of FAME in mm^2/s. The density at 20 °C of biodiesel was acquired from

$$\rho = \mathbf{0.8463} + ((\mathbf{4.9})/\boldsymbol{M}) + (\mathbf{0.0118}).(\boldsymbol{N}) \tag{4}$$

where ρ is the density of FAME in g/cm^3. The calorific value of biodiesel was calculated from

$$\boldsymbol{\delta = 46.19 - 1794/M - (0.21).(N)} \tag{5}$$

where δ is the calorific value of the fatty acid (FAME) in MJ/kg.

3 Biodiesel Derived from Multiple Feedstocks of Palm Oil, Coconut Oil, and Corn Oil

3.1 *The Selection of Multiple Vegetable Oil Feedstock*

The selection of ratio for the oil mixture at 5:4:1 (corn oil: palm oil: coconut oil) on the basis of the respective biodiesel's fuel properties (Table 1). The ratio of corn oil was the largest because of its excellent fuel properties including calorific value, cetane number, cloud point and flash point, relative to palm oil and coconut oil.

Transesterification in this scenario was performed on a mixture of vegetable oil from coconut, corn, and palm oil by varying the reaction temperatures (50, 60 and 70 °C) catalyzed by KOH. Table 2 depicts that transesterification at 70 °C resulted on the highest yield biodiesel from vegetable oil, followed by 60 and 50 °C. It can be deduced that higher the reaction temperature of transesterification process leads to a higher yield of FAME from vegetable oil. This finding was in agreement with the study of Demirbas (2009b) on biodiesel synthesis from waste cooking oil catalyzed by KOH, where increasing reaction temperature increased biodiesel yield (Demirbas 2009b). Higher reaction temperature also yielded a higher composition of biodiesel (FAME) from crude biodiesel, and lower composition of by-products, which is glycerol.

Table 1 Properties of biodiesel standard and biodiesel from coconut oil, corn oil, and palm oil (Sajjadi et al. 2016; Singh & Singh 2010)

Fuel properties	Coconut biodiesel	Corn biodiesel	Palm biodiesel
Calorific value (MJ/kg)	35.2	39.9	33.5
Cetane number	64.65	46.65	62
Cloud point (°C)	−1.6	−3	13
Density (kg/m^3)	867	883	880
Flash point (°C)	75	277	164
Kinematic viscosity at 40 °C (mm^2/s)	3.20	4.19	5.7

Table 2 Biodiesel yield from transesterification of multiple vegetable oils feedstock (corn oil: palm oil: coconut oil, 5:4:1) at varying transesterification reaction temperature

Chemical composition of biodiesel	Temperature for transesterification (°C)		
	50	60	70
FAME yield (%, w/w)	75.04	78.17	79.73
FAME composition in crude biodiesel (%, w/w)	78.39	79.52	80.44
Glycerol composition in crude biodiesel (%, w/w)	21.61	20.48	19.56

3.2 *Chemical Bonds and Chemical Functional Groups of Biodiesel Derived from Palm Oil, Coconut Oil, and Corn Oil*

FTIR spectra for three different samples of biodiesel prepared from transesterification at different temperature are shown in Fig. 2 in order to assess the functional groups present in FAME. The spectra of three samples showed subtle differences of peak values as different reaction temperature did not affect the functional groups of the compounds in FAME samples.

The peak at ~723 cm^{-1}, overlapping with =C–H groups, indicated vibration rocking mode. These groups indicated the functional methylene group within FAME ($-(CH_2)_n-$) (Oyerinde & Bello 2016). The vibration rocking mode is an indicator that FAME composed of the aliphatic structure of the long chain (Oyerinde & Bello 2016). The peak characteristics found in the 1070–1250 cm^{-1} regions suggested stretching mode of vibrations of both C–O–C and C–O, that indicated O–CH_3's bending vibration (Mohammed et al. 2015). The band area of 1377–1465 cm^{-1} could be related to C–H methyl group's bending vibration (Mohammed et al. 2015). The peak characteristics at wavenumber 1742 cm^{-1} which are due to the stretching vibration of C=O groups, which suggested the presence of carbonyl groups in the esters of FAME (Leiva-Candia et al. 2015). The peaks of C-H alkane groups at ~2854 cm^{-1} and ~2925 cm^{-1} indicated asymmetric and symmetric stretching vibrations, respectively of methyl (CH_3) or methylene (CH_2) groups in the biodiesel ester chains that require substantial energy to instigate stretching vibrations in its bond relative to the common C–H bending vibrations of low-energy and frequency region detected in alkene groups (Oyerinde & Bello 2016). At ~3009 cm^{-1} the peak is due to the stretching vibration of groups of =C–H alkene, possibly attributed to the unsaturated fatty acid that presented in FAME (Leiva-Candia et al. 2015).

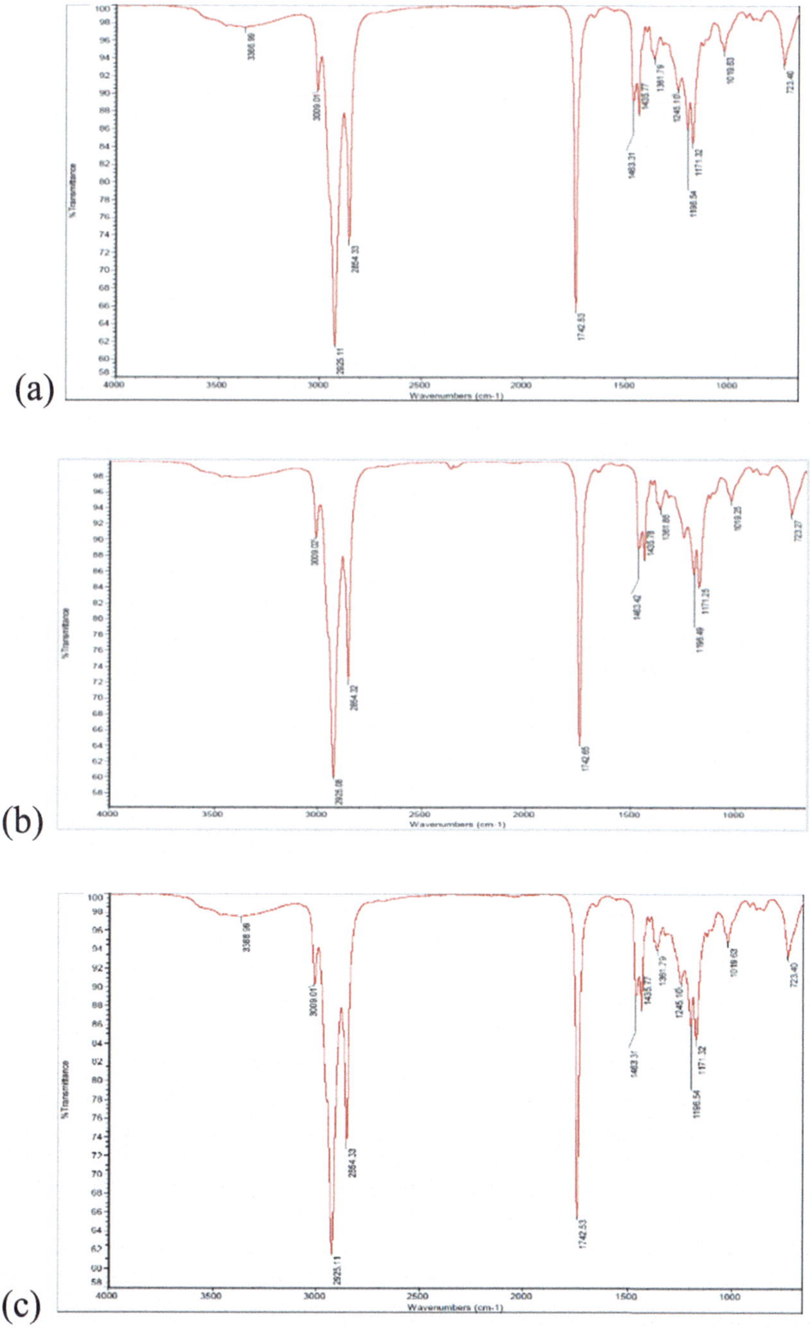

Fig. 2 FTIR spectra of biodiesel synthesized via transesterification of multiple vegetable oils feedstock (corn oil: palm oil: coconut oil, 5:4:1) at varying transesterification reaction temperature (**a** 50 °C, **b** 60 °C, **c** 70 °C)

3.3 *FAME Analysis of Biodiesel Derived from Palm Oil, Coconut Oil, and Corn Oil*

Biodiesel prepared at different reaction temperature was analyzed via GC–MS during transesterification to assess the composition of FAME in all biodiesel samples (Fig. 3). All biodiesel consisted of major fatty acids that typically present in vegetable oil-derived biodiesel, which were palmitic acid, stearic acid, oleic acid, linoleic acid, and linolenic acid, that refers to the number of carbon and bonds of C16:0, C18:0, C18:1, C18:2, and C18:3 respectively (Ahmad et al. 2016).

Biodiesel prepared at 70 °C was shown to have the highest composition of saturated and monounsaturated FAME at 96.17% (Table 3), which is the most desirable composition of fatty acid for biodiesel (Ahmad et al. 2015). The presence of PUFA in biodiesel should be minimal as it leads to poor fuel oxidative stability during storage (Ahmad et al. 2015; Chisti 2007).

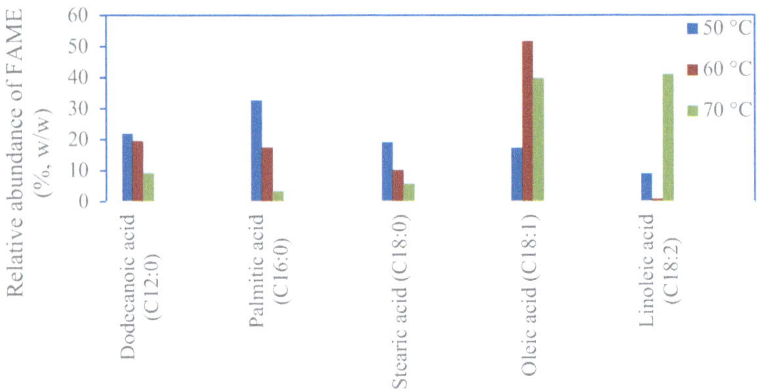

Fig. 3 Fatty acids composition of FAME or biodiesel from transesterification of multiple vegetable oil feedstock (corn oil: palm oil: coconut oil, 5:4:1) at varying transesterification reaction temperature (50–70 °C)

Table 3 Composition (% w/w) of saturated, monounsaturated, and polyunsaturated fatty acids (SFA, MUFA, and PUFA respectively) of biodiesel prepared at varying reaction temperature

Types of saturation of fatty acids	Transesterification reaction temperature (°C)		
	50	60	70
SFA	68.60	15.24	66.11
MUFA	18.47	35.11	30.06
PUFA	12.93	49.66	3.83
Unsaturated fatty acid	31.40	84.76	33.89

3.4 Fuel Properties of Biodiesel Derived from Palm Oil, Coconut Oil, and Corn Oil

Fuel properties of biodiesel were evaluated empirically based on FAME composition (Table 4). Calorific value of fuels is a critical indicator of how much energy is produced for generating work (Miri et al. 2017). Calorific value of biodiesel is therefore can be attributed to the reduction in engine capacity (Madiwale et al. 2017). The heat content for methyl esters of biodiesel elevated as the chain length of the fatty acids decreases (Lujaji et al. 2010). Biodiesel samples from this study have calorific value comparable to diesel. Different transesterification temperatures have negligible influence on the calorific value of biodiesel. Biodiesel prepared from transesterification reaction at 70 °C has the highest cetane number relative to other biodiesel samples. High values of cetane number are desirable as it can enhance the combustion that may result in low exhaust emissions, minimal noise and excellent engine performance (Madiwale et al. 2017). The cetane number of biodiesel from multiple feedstocks was greater than diesel and substantially higher than corn biodiesel. The values of kinematic viscosity of the FAME samples were within the range of 1.9–6.0 (mm^2)/s as proposed by ASTM biodiesel standard. Kinematic viscosity of biodiesel from multiple feedstocks was also lower than that of palm biodiesel. High viscosity could lead to issues with fuel flow and cause failure in the fuel pump while low viscosity could increase engine wear and leakage (Oyerinde & Bello 2016). Density is another critical fuel property that significantly impacts engine performance as it impacts the atomization efficiency of the fuel and the combustion characteristics (Flood et al. 2016). Density of each biodiesel samples in this study was slightly greater than diesel fuel and within EN 14214 biodiesel standard of 860–900 kg/m^3 (Sajjadi et al. 2016).

Table 4 Fuel properties comparison of FAME sample prepared from transesterification of multiple vegetable oils feedstock (corn oil: palm oil: coconut oil, 5:4:1) at varying transesterification reaction temperature to biodiesel (Sajjadi et al. 2016; Singh & Singh 2010) and petrodiesel (Singh & Singh 2010)

Fuel properties	Biodiesel from multiple feedstock (varying reaction temperature)			Biodiesel (ASTM standard or EN 14214)	Diesel
	50 °C	60 °C	70 °C		
Calorific value, δ (MJ/kg)	39.59	39.43	39.61	34.4–45.2	43.80
Cetane number, Ø	58.13	56.72	61.68	48–65	50
Density, ρ (kg/m^3)	875.5	875.9	873.9	860–900	855
Kinematic viscosity, φ (mm^2/s)	4.13	3.99	4.42	1.9–6	3.06

4 Conclusion

Biodiesel or FAME was produced from transesterification of the combination of coconut oil, corn oil, and palm oil with methanol at varying reaction temperature (50–70 °C) and KOH as the catalyst. Biodiesel produced from the multiple feedstocks showed improved performance in some fuel properties in comparison to biodiesel prepared from single vegetable oil feedstock, especially for cetane number values and kinematic viscosity. The cetane number values of biodiesel from multiple vegetable oil feedstock from this study were even higher than that of diesel. Therefore, the use of multiple feedstock vegetable oil in biodiesel production could improve the fuel properties of biodiesel from single vegetable oil feedstock.

Acknowledgements Authors acknowledge financial support from the Ministry of Higher Education Malaysia through Fundamental Research Grant Scheme (FRGS) (FRGS/1/2019/TK02/UIAM/02/3). The authors would also like to acknowledge the Ministry of Higher Education, Malaysia, for the Postdoctoral Fellowship scheme of the corresponding author.

References

Ahmad, F. B., Zhang, Z., Doherty, W. O. S., & O'Hara, I. M. (2015). A multi-criteria analysis approach for ranking and selection of microorganisms for the production of oils for biodiesel production. *Bioresource Technology, 190*, 264–273. https://doi.org/10.1016/j.biortech.2015.04.083

Ahmad, F. B., Zhang, Z., Doherty, W. O. S., & O'Hara, I. M. (2016). Evaluation of oil production from oil palm empty fruit bunch by oleaginous micro-organisms. *Biofuels, Bioproducts and Biorefining, 10*(4), 378–392. https://doi.org/10.1002/bbb.1645

Ahmad, F. B., Zhang, Z., Doherty, W. O., & O'Hara, I. M. (2018). Optimising extraction of microalgal oil using accelerated solvent extraction by response surface methodology. *Journal of Engineering Science and Technology, 13*(4), 964–976.

Ahmad, F. B., Zhang, Z., Doherty, W. O. S., & O'Hara, I. M. (2019). The prospect of microbial oil production and applications from oil palm biomass. *Biochemical Engineering Journal, 143*, 9–23. https://doi.org/10.1016/j.bej.2018.12.003

Annamalai, R., Kuppusamy, R., & Krishnan, R. S. (2018). Effect of piston bowl geometry and different injection pressure on the performance, emission and combustion characteristics of diesel engine using biodiesel blend. *Thermal Science, 22*(3), 1445–1456.

Anwar, M., Rasul, M. G., & Ashwath, N. (2019). The efficacy of multiple-criteria design matrix for biodiesel feedstock selection. *Energy Conversion and Management, 198*, 111790. https://doi.org/10.1016/j.enconman.2019.111790

Bongfa, B., Hamid, M. K. A., Samin, P. M., & Adeoti, M. O. (2020). Coconut-castor oil mixture: The potential of flow to tribo-contacts of internal combustion engine at low-temperatures. *Jurnal Tribologi, 27*, 132–142.

Chiou, B. S., El-Mashad, H. M., Avena-Bustillos, R. J., Dunn, R. O., Bechtel, P. J., McHugh, T. H., Imam, H. S., Glenn, G. M., Orts, W. J., & Zhang, R. (2008). Biodiesel from waste salmon oil. *Transactions of the ASABE, 51*(3):797–802. https://doi.org/10.13031/2013.24517

Chisti, Y. (2007). Biodiesel from microalgae. *Biotechnology Advances, 25*(3), 294–306. https://doi.org/10.1016/j.biotechadv.2007.02.001

Cooperation, A. P. E. (2009). *Establishment of the guidelines for the development of biodiesel standards in the APEC region*. APEC.
Demirbas, A. (2009a). Progress and recent trends in biodiesel fuels. *Energy Conversion and Management, 50*(1), 14–34. https://doi.org/10.1016/j.enconman.2008.09.001
Demirbas, A. (2009b). Biodiesel from waste cooking oil via base-catalytic and supercritical methanol transesterification. *Energy Conversion and Management, 50*(4), 923–927.
Flood, M. E., Connolly, M. P., Comiskey, M. C., & Hupp, A. M. (2016). Evaluation of single and multi-feedstock biodiesel–diesel blends using GCMS and chemometric methods. *Fuel, 186*, 58–67.
Habibullah, M., Masjuki, H. H., Kalam, M. A., Rizwanul Fattah, I. M., Ashraful, A. M., & Mobarak, H. M. (2014). Biodiesel production and performance evaluation of coconut, palm and their combined blend with diesel in a single-cylinder diesel engine. *Energy Conversion and Management, 87*, 250–257. https://doi.org/10.1016/j.enconman.2014.07.006
Hazmi, A. T., Ahmad, F. B., Athoillah, A. Z., & Jameel, A. T. (2021). Comparison of ex-situ and in-situ transesterification for the production of microbial biodiesel. *Bulletin of Chemical Reaction Engineering & Catalysis*, 11. https://doi.org/10.9767/bcrec.16.4.11044.733-743
Leiva-Candia, D., Tsakona, S., Kopsahelis, N., Garcia, I., Papanikolaou, S., Dorado, M., & Koutinas, A. (2015). Biorefining of by-product streams from sunflower-based biodiesel production plants for integrated synthesis of microbial oil and value-added co-products. *Bioresource Technology, 190*, 57–65.
Lujaji, F., Bereczky, A., Janosi, L., Novak, C., & Mbarawa, M. (2010). Cetane number and thermal properties of vegetable oil, biodiesel, 1-butanol and diesel blends. *Journal of Thermal Analysis and Calorimetry, 102*(3), 1175–1181.
Madiwale, S., Karthikeyan, A., & Bhojwani, V. (2017). A comprehensive review of effect of biodiesel additives on properties, performance, and emission. In *IOP conference series: Materials science and engineering* (Vol. 197, p. 012015). https://doi.org/10.1088/1757-899x/197/1/012015
Manigandan, S., Gunasekar, P., Devipriya, J., & Nithya, S. (2019). Emission and injection characteristics of corn biodiesel blends in diesel engine. *Fuel, 235*, 723–735.
Martani, F., Maestroni, L., Torchio, M., Ami, D., Natalello, A., Lotti, M., Porro, D., & Branduardi, P. (2020). Conversion of sugar beet residues into lipids by Lipomyces starkeyi for biodiesel production. *Microbial Cell Factories, 19*(1), 204. https://doi.org/10.1186/s12934-020-01467-1
Miri, S. M. R., Seyedi, S. R. M., & Ghobadian, B. (2017). Effects of biodiesel fuel synthesized from non-edible rapeseed oil on performance and emission variables of diesel engines. *Journal of Cleaner Production, 142*, 3798–3808.
Mohammed, I. Y., Abakr, Y. A., Kazi, F. K., Yusup, S., Alshareef, I., & Chin, S. A. (2015). Comprehensive characterization of Napier grass as a feedstock for thermochemical conversion. *Energies, 8*(5), 3403–3417.
Oyerinde, A., & Bello, E. (2016). Use of fourier transformation infrared (FTIR) spectroscopy for analysis of functional groups in peanut oil biodiesel and its blends. *Current Journal of Applied Science and Technology*, 1–14.
Pontoh, J. (2016). Gas chromatographic analysis of medium chain fatty acids in coconut oil. *The Journal of Pure and Applied Chemistry Research, 5*(3), 157.
Ramírez-Verduzco, L. F., Rodríguez-Rodríguez, J. E., & Jaramillo-Jacob, Ad. R. (2012). Predicting cetane number, kinematic viscosity, density and higher heating value of biodiesel from its fatty acid methyl ester composition. *Fuel, 91*(1), 102–111. https://doi.org/10.1016/j.fuel.2011.06.070
Sajjadi, B., Raman, A. A. A., & Arandiyan, H. (2016). A comprehensive review on properties of edible and non-edible vegetable oil-based biodiesel: Composition, specifications and prediction models. *Renewable and Sustainable Energy Reviews, 63*, 62–92. https://doi.org/10.1016/j.rser.2016.05.035
Samsukumar, R., Muaralidhararao, M., Krishna, A. G., Jayaraju, Y., Vatsav, P. S. S., Manikanta, V., Satish, S., & Bhuvannaidu, S. (2015). Performance and emission analysis on CI engine with

palm oil biodiesel blends at different fuel injection pressures. *International Journal of Innovative Science and Research Technology, 4*(4), 2516–2527.

Singh, S., & Singh, D. (2010). Biodiesel production through the use of different sources and characterization of oils and their esters as the substitute of diesel: A review. *Renewable and Sustainable Energy Reviews, 14*(1), 200–216.

Sudrajat, A., Tamaldin, N., Yamin, A. K. M., Bin, M. F., Tunggal, D., & Chichiri, B. (2020). Performance analysis of biodiesel engine by addition of HHO gas as a secondary fuel. *Jurnal Tribologi, 26*, 120–134.

Zicker, M. C., Craig, A. P., de Oliveira, R. D., Franca, A. S., Labanca, R. A., & Ferreira, A. V. M. (2016). Quantitative analysis of acidity level in virgin coconut oils by Fourier transform infrared spectroscopy and chemometrics. *European Journal of Lipid Science and Technology, 118*(9), 1350–1357.

Pretreatment of Sago Waste for Biohydrogen Production: Towards a Sustainable Circular Bioeconomy

Noor Illi Mohamad Puad, Tami Astie Ulhiza, Mohd Izhar Abdul Malek, Nurainin Farhan Abd Rahim, and Ummul Madihah Mohd Rashidi

Abstract Hydrogen is a promising future energy source because it has a high energy content and produces the least CO_2. It contains 122 kJ/g of energy, which is 2.75 times more than hydrocarbon fuels. Biohydrogen production via dark fermentation offers simple mechanisms compared to other biohydrogen production routes, such as photofermentation and hybrid processes. By utilising waste as a substrate, microorganisms can convert sugars into hydrogen. Dark fermentative bacteria not only produce hydrogen but also reduce toxins in the wastewater. However, pretreatment of waste as the substrate for dark fermentation poses challenges. The process must break down complex sugars into fermentable ones without prohibitively expensive or releasing harmful chemicals. This chapter explores various pretreatment methods for sago waste in biohydrogen production.

Keywords Biohydrogen · Dark fermentation · Sago waste · Pretreatment

1 Introduction

Hydrogen has gained global attention as an energy carrier for an alternative clean fuel in the future by which it is able to produce water as a by-product (Qazi 2022). Moreover, as compared to other fuels, hydrogen carries the highest energy per unit mass and the lowest CO_2 content (McCay et al. 2020). Kapdan and Kargi (2006) reported that hydrogen has a 2.75 times higher energy yield than hydrocarbon fuels. However, conventional hydrogen production technology still carries some drawbacks from economic and environmental viewpoints. It requires extra energy and excessive cost to produce hydrogen energy (Das et al. 2014; Osman et al. 2022). In addition, the production process by conventional methods still releases pollution into the atmosphere.

N. I. M. Puad (✉) · T. A. Ulhiza · M. I. A. Malek · N. F. A. Rahim · U. M. M. Rashidi
Department of Chemical Engineering and Sustainability, Faculty of Engineering, International Islamic University Malaysia, Kuala Lumpur, Malaysia
e-mail: illi@iium.edu.my

F. Adyani Ahmad Fuad et al. (eds.), *Circular Bioeconomy*,
https://doi.org/10.1007/978-981-97-7010-6_6

Thus, to overcome this, researchers have found alternatives to harvest hydrogen using the least energy. Production of hydrogen by biological means has attracted the attention of researchers to investigate its potential for an inexhaustible, low-cost, and renewable source of clean energy (Das et al. 2014). Biohydrogen production involves a variety of microorganisms. Typically, it involves cyanobacteria and algae in indirect and direct biophotolysis. On the other hand, bacteria carry out the fermentation process, which can be classified as dark fermentation, photofermentation, or a hybrid process.

Dark fermentation shows a simpler mechanism and preparation than photofermentation and hybrid processes. Dark fermentation is an easy process for large production in the future. In the dark fermentation, hydrogen is produced without the assistance of light. Most of the dark adaptive microorganisms are usually found in those places where the oxygen is devoid, for example, under the soil, wastewater, lake, river, ocean, sludge, and sewage, and inside the vegetable (Das et al. 2014). The microorganism used in the dark fermentation varies, including mixed cultures obtained from the anaerobic sludge digesters, co-culture, and single cultures of known species of hydrogen-producing bacteria (Soares et al. 2020).

Initially, the research for biohydrogen production utilized substrate from laboratory-grade chemicals such as glucose, glycerol, acetate, and butyrate, which require a high production cost. Even though the yield is high, using synthetic chemicals is not economical for the continuous production of biohydrogen. On the other hand, waste is considered a cheap option for the sustainability of raw materials. Waste material is more economical as well as environmentally friendly. In addition, according to Mishra and Das (2014) who utilized distillery effluent, generating biohydrogen from waste not only helps in getting biohydrogen but also helps in the bioremediation of wastewater. Rice bran, crude glycerol, apple pomace, cassava, dairy and citrus processing wastewaters, and sago starch effluent are reported to yield significant amounts of hydrogen during dark fermentation (Ulhiza et al. 2017). Those wastes contain sugar which can be converted to hydrogen. Dark fermentative bacteria are not only able to produce hydrogen but also reduce toxins in the wastewater.

Several aspects including the availability, cost, carbohydrate content, and biodegradability must be considered in the selection of waste for biohydrogen production (Mogili et al. 2022). Although there are many reports on the process of technology in wastewater, only a few focus on the pretreatment process. Indeed, one of the challenges in the production of hydrogen from waste is the pretreatment (Urbaniec & Bakker 2015). The pretreatment process should be able to break down as many complex sugars as possible into fermentable sugar. At the same time, the method should not be too expensive, as it will affect the overall total cost of production. Moreover, it should not release toxic chemicals that can harm the bacteria as the biohydrogen producer.

This chapter will highlight the various pretreatment methods of lignocellulosic waste for biohydrogen production and discuss different pretreatment approaches to sago waste to produce fermentable sugars for biohydrogen production.

2 Pretreatment of Lignocellulosic Waste

Pretreatment involves the treatment of wastes prepared for digestion, which results in delignification, breaking down the cellulose, hemicellulose, and reducing the degree of polymerization, which increases the hydrolysis rate (Bensah & Mensah 2013). Different methods for pretreatment include physical, chemical, biological, and physicochemical methods (Baruah et al. 2018). Among other pretreatment methods, physical and chemical approaches are widely being used for processing lignocellulosic materials (Baruah et al. 2018; Ethaib et al. 2015).

Physical pretreatment involves the breakdown of biomass size and structure by grinding or milling. This method can increase the accessibility of the materials and decrease the crystallinity and degrees of polymerization of cellulose (Taherzadeh & Karimi 2008). The energy required for physical pretreatment depends on the size and the crystallinity of lignocellulosic materials. Irradiation is also classified as physical pretreatment since it could change the ultrastructure of cellulose. Examples of irradiation are gamma-ray irradiation, electron-beam irradiation, and microwave irradiation. Among other irradiation, microwave irradiation is usually chosen due to its high heating efficiency and easy operation (Ethaib et al. 2015). Microwave heating was used in the degradation of various types of lignocellulosic materials such as wheat bran, barley husk, bamboo sawdust, Miscanthus sp., and sugarcane leaves (Baruah et al. 2018). Microwave heating is effective because it supplies electromagnetic waves with different frequencies that give a direct interaction between polar molecules and electromagnetic fields (Zhu et al. 2015).

In most pretreatment processes, acid is preferred due to its ability to remove the hemicellulose portion in liquid fraction, especially in wastewater treatment. Acid pretreatment will result in the disruption of covalent bonds, hydrogen bonds, and van der Waals interaction that hold the biomass structure (Amin et al. 2017). This will cause the solubilization of hemicellulose, which will increase biomass accessibility. In acid pretreatment, it is vital to identify the types of acid to be used, either concentrated or diluted acid. However, concentrated acid is rarely used in pretreatment due to its hazardous effect and high corrosiveness to the equipment. In addition, concentrated acid will result in the biomass structure being ineffectively degraded. This is because the reaction with concentrated acid requires mild temperature and longer reaction time, which results in minimum sugar degradation. Therefore, among other acids, dilute H_2SO_4 is preferred because it is cheaper and yields a better reducing or fermentable sugars.

Alkaline pretreatment is performed by mixing the lignocellulosic biomass with a strong or dilute base. Examples of strong bases are sodium hydroxide (NaOH), calcium hydroxide ($Ca(OH)_2$), and potassium hydroxide (KOH), while weak base is ammonia (NH_3). This method is able to degrade the ester and glycosidic chain of biomass, leading to lignin structure disruption (Kim et al. 2016). Besides that, cellulose crystallinity and rate of polymerization can be reduced. However, the effect of pretreatment depends on the lignin content in which different biomass will give different pretreatment results. Even though alkaline pretreatment can be performed at low temperatures, it is reported that alkaline pretreatment is the second option after acid pretreatment due to the high reaction time required with a high concentration of base needed (Bensah & Mensah 2013; Taherzadeh & Karimi 2008).

Biological pretreatment involves the use of microorganisms to pretreat lignocellulosic materials. This method is environmentally friendly, saves energy, and does not require chemicals for pretreatment. However, it takes a longer time, and the rate of hydrolysis is low compared with other pretreatment approaches. The bacteria will degrade hemicellulose and lignin but very little cellulose. This is because cellulose is more resistant to biological attack than other parts of lignocelluloses (Vasco-Correa et al. 2016). Microorganisms suitable for this purpose are brown-, white- and soft-rot fungi. Lignin degradation of white rot fungi is the most effective since the reaction involves lignin-degrading enzymes such as laccase (Janusz et al. 2017). However, the high price of enzymes has caused this method to be unfavorable for industrial purposes.

Besides that, another alternative for pretreatment is physicochemical methods. Physicochemical pretreatment combines physical and chemical techniques to break down the structure of lignocellulosic materials. In this method, biomass is treated at high temperatures, pressure, and high concentrations of acid or alkaline. This process will lead to the degradation of hemicellulose and lignin, which it increases the potential of cellulose hydrolysis (Baruah et al. 2018).

3 Sago Waste as Biohydrogen Feedstock

Sago palm (*Metroxylon sp.*) is a starchy-based crop found abundantly in tropical countries. Sago palm is widely planted in Malaysia. The plantation has been well established and become one of the sources of national income by export. Malaysian's annual sago export is around 47,000 mt (Jong et al. 2018). Sago palm grows to a height of 10–12 and 0.8–1 m in diameter. The plant is hapaxanth, which is flowering only once in a lifetime and then dying subsequently (Amin et al. 2019). Once it dies, the new plant will regrow from the sucker. Thus, replanting sago palms is unnecessary.

The sago starch industry yields sago flour as the main product. During this process, three major by-products, which are sago trunk, fibrous pith residues, and wastewater, are generated. Annually, it is reported that 2.5 million tons of effluent are generated from this industry (Yunus et al. 2014). Yet sago wastewater is usually left untreated. Untreated sago wastewater contains a high concentration of COD and organic acid. Therefore, it is essential to neutralize the waste before discharging it to the environment. The digestion effluent of sago wastewater could reduce COD, ammoniacal-nitrogen, and phosphate to the levels of 98.0, 99.9 and 99.4%, respectively (Awg-Adeni et al. 2010).

The fermentation of sago wastewater is useful for biogas and bioethanol industry (Yunus et al. 2014). Previous studies have primarily concentrated on the composition of sago wastewater. Sugar from sago wastewater is usually found as a polysaccharide.

Table 1 Characteristics of sago wastewater in three locations

Parameter	Ulhiza et al. (2018)	Yunus et al. (2014)	Anbukumar and Kumar (2014)
pH	3.74	4.2	5.2
COD (mg/L)	23,700	10,640	2,080
TDS (mg/L)	1,017	NR	2,010
TSS (mg/L)	14,000	1,998	86
VSS (mg/L)	39.5	NR	NR
Total carbohydrate (mg/L)	2,781	557	NR

* NR: Not reported

Glucose content in the untreated sago wastewater was reported to be 1.153 g/L (Puad et al. 2015). Yunus et al. (2014) characterized and compared the composition of sago starch processing effluent or sago wastewater collected from the sago mill in Sarawak and the one that was processed under laboratory conditions. It was observed that the total pH, COD, total solid, total suspended solids, and soluble carbohydrates from the latter were 4.2, 10.64 g/L, 6047 mg/L, 1998 mg/L, and 557 mg/L, respectively. In addition, trace elements (P, K, Ca, Mg, Fe, Na) and minor elements (Zn, Mn, Cu, Ni) were also detected, but in a small amount.

Meanwhile, Ulhiza et al. (2018) reported the characteristics of sago wastewater collected from a sago mill in Batu Pahat, Johor, Malaysia. The sago wastewater was acidic in nature, with a low pH level of 3.63. In addition, it contained more solids compared to the one reported for a sago mill in Sarawak, Malaysia (Yunus et al. 2014) and Tamil Nadu, India (Anbukumar & Kumar 2014) (Table 1). TSS, TDS, and VSS are the measurements used to quantify the solids in the wastewater. TSS refers to a solid that cannot pass through a filter paper and is usually retained in the filter paper. TDS is a solid dissolved in wastewater. Meanwhile, VSS is a solid lost during ignition (heating to 550 °C). COD measures the oxygen content in the wastewater subjected to oxidation by strong chemical oxidants (for an example, dichromate). COD is an important method to approximate organic matter present in wastewater. COD in Ulhiza et al. (2018) was found to be 23,700 mg/L, which is higher than Yunus et al. (2014) i.e. 10,640 mg/L). This indicates that many organic compounds are found in the sediment of sago wastewater. Microorganisms could use these compounds as a carbon source during hydrogen production.

4 Pretreatment of Sago Waste

The sago starch process discharges a high amount of sugar-containing waste. However, the sugars are not readily used for fermentation. The pretreatment method is essential to reduce sugars into fermentable sugars and eliminate other substances that might be toxic to the bacteria. The common hydrolysis methods in sago waste pretreatment are acid and enzymatic pretreatment (Table 2). In acid hydrolysis, acid acts as a catalyst that breaks the glycosidic bond into maltose, maltotriose, glucose, and dextrin, depending on which chain is attacked (Kumoro et al. 2008). While in the enzymatic hydrolysis, amylase is used to breakdown the starch into fermentable sugars (Sunaryanto et al. 2013).

In a preliminary study of hydrogen production from sago wastewater by *E. coli*, Puad et al. (2015) hydrolysed starch by the acid hydrolysis method. Sago wastewater was initially filtered to remove solids. Subsequently, it was added with 0.5 M of sulphuric acid and exposed to heat at 90 °C. After neutralisation by 0.5 M NaOH, the sample was set on ice to absorb the high temperature. Glucose content after pretreatment was reported to increase by 86% from 1.153 to 2.150 g/L.

Table 2 Pretreatment method used to hydrolyse sago waste

Pretreatment method	Pretreatment condition (s)	Sugar content before pretreatment	Sugar content after pretreatment	References
Enzymatic hydrolysis	1% α-amylase and γ-amylase	0.5–0.7 g/L soluble carbohydrate	6 g/L soluble carbohydrate	Yunus et al., (2014)
Enzymatic hydrolysis	1% α-amylase	0.5–0.7 g/L soluble carbohydrate	3 g/L soluble carbohydrate	Yunus et al., (2014)
Enzymatic hydrolysis	5% (α-amylase and γ-amylase	0.5–0.7 g/L soluble carbohydrate	24 g/L	Yunus et al., (2014)
Enzymatic hydrolysis	0.1 ml 8222 U mL and α-amylase	*ND	7% (w/v) fermentable sugar	Sunaryanto et al., (2013)
Enzymatic hydrolysis	Glucoamylase 6 AGU/ml	*ND	0.56 g glucose/g waste	Kumoro et al., (2008)
Acid hydrolysis	1 M H_2SO_4 at 100°C for 60 min	*ND	9.138 g/L	Mohamad Puad et al., (2018)
Acid hydrolysis	2.5% H_2SO_4 for at 121°C 120 min	*ND	6.6% (w/v) fermentable sugar	Sunaryanto et al., (2013)
Acid hydrolysis	1.5 M H_2SO_4 90 °C for 120 min	*ND	0.62 g glucose/g waste	Kumoro et al., (2008)
Acid hydrolysis	0.5 M H_2SO_4 90 °C for 40 min	1.153 g/L glucose	2.15 g/L glucose	Puad et al., (2015)
Microwave-assisted acid hydrolysis	0.6 M of H_2SO_4 for 9 min (100 W)	Not available	3.920 g/L	Puad et al., (2020)

* ND = Not determined

As a comparison, Yunus et al. (2014) investigated the effect of enzymatic hydrolysis on the soluble carbohydrate content in the sago wastewater. Before the pretreatment, soluble carbohydrate was 557 mg/L. The addition of 1% of α-amylase enhanced the soluble carbohydrate content to 3000 mg/L. The content was doubled further when a combination of 1% of α-amylase and γ-amylase was used concomitantly. The increment could be attributed to the ability of γ-amylase to cleave α-1,4- and α-1,6-glycosidic linkages in polysaccharides, dextrins, and sugars. This finding was consistent with the previous finding (Woiciechowski et al. 2002) which performed an experimental investigation on enzymatic hydrolysis to recover fermentable sugar from cassava bagasse. This method effectively converts starch into amylose, a mostly linear α-D—(1–4)-glucan, and branched amylopectin to fermentable sugars.

Basically, microwave-assisted acid pretreatment has the potential to produce a high concentration of glucose in a short period of time (Table 2). According to Puad et al. (2020), it was observed that the maximum glucose obtained was 3.920 g/L in only 9 min of residence time. This value can be compared with the conventional acid pretreatment method. It was reported that the highest glucose concentration achieved was 9.13 g/L at 100 °C reaction temperature for 60 min of incubation time, which is comparatively higher than in microwave-assisted acid pretreatment (Mohamad Puad et al. 2018). It showed that even though the conventional acid pretreatment led to high glucose concentration, a longer pretreatment time and high reaction temperature were needed. This agrees with other acid pretreatment reported by Kumoro et al. (2008), Sunaryanto et al. (2013), and Puad et al. (2015).

So far, there has been limited discussion about the drawbacks of each pretreatment method. Even though enzymatic hydrolysis seems more efficient, purchasing the enzyme is expensive. Using enzymes for starch hydrolysis on a large scale is probably not feasible. The optimization of the conditions to enhance fermentable sugar becomes very important. On the other hand, acid hydrolysis creates harsh conditions for microorganisms. A pH that is too low and a high temperature is not preferred for living bacteria. Therefore, neutralisation and reduction of the temperature before the fermentation process must be considered. Other than enzymatic and acid hydrolysis, pretreatment methods for starch include ultrasonication, alkali hydrolysis, and combined hydrolysis (acid and enzyme) (Ramprakash & Muthukumar 2015). A more advanced pretreatment method found that 43.8% of glucose yield was recovered from the microwave hydrothermal hydrolysis of sago pith waste (Thangavelu et al. 2014). Even though microwave is an effective hydrolysis method in terms of increasing energy efficiency, reducing processing times, avoiding the generation of degradation products, and avoiding superficial overheating, it still requires extensive research on dielectric properties and non-thermal effect and microwave engineering (Aguilar-Reynosa et al. 2017).

5 Conclusion

Sago waste is a potential feedstock for biohydrogen production. This chapter highlights the importance of pretreatment methods in breaking down complex sugars into fermentable ones. While various methods, such as physical, chemical, biological, and physicochemical, are explored, each presents its own challenges and benefits. For instance, while enzymatic hydrolysis is efficient, the cost of enzymes makes it less feasible for large-scale operations. On the other hand, acid hydrolysis creates harsh conditions for microorganisms, necessitating additional steps for neutralisation and temperature reduction. Further research, particularly in microwave engineering, is recommended to optimise these processes and enhance biohydrogen production from sago waste.

Acknowledgements The authors acknowledge financial support from the IIUM Research Initiative Grant (RIGS16-089-0253).

References

Aguilar-Reynosa, A., Romaní, A., Ma Rodríguez-Jasso, R., Aguilar, C. N., Garrote, G., & Ruiz, H. A. (2017). Microwave heating processing as alternative of pretreatment in second-generation biorefinery: An overview. *Energy Conversion and Management, 136*, 50–65. https://doi.org/10.1016/j.enconman.2017.01.004

Amin, F. R., Khalid, H., Zhang, H., Rahman, S. U., Zhang, R., Liu, G., et al. (2017). Pretreatment methods of lignocellulosic biomass for anaerobic digestion. *AMB Express, 7*(1), 72. https://doi.org/10.1186/s13568-017-0375-4

Anbukumar, S., & Kumar, A. (2014). Effluent treatment for sago industry using zeolite and activated carbon. *Open Journal of Water Pollution and Treatment, 2*, 18–26. https://doi.org/10.15764/WPT.2014.02003

Awg-Adeni, D., Abd-Aziz, S., Bujang, K., & Hassan, M. (2010). Bioconversion of sago residue into value added products, *9*, 2016–2021.

Baruah, J., Nath, B., Sharma, R., Kumar, S., Deka, R., Baruah, D., et al. (2018). Recent trends in the pretreatment of lignocellulosic biomass for value-added products. *Front Energy Research, 6*(141). https://doi.org/10.3389/fenrg.2018.00141

Bensah, E. C., & Mensah, M. (2013). Chemical pretreatment methods for the production of cellulosic ethanol: Technologies and innovations. *International Journal of Chemical Engineering, 2013*, 719607. https://doi.org/10.1155/2013/719607

Das, D., Khanna, N., & Dasgupta, C. N. (2014). *Biohydrogen production: Fundamentals and technology advances*. CRC Press.

Ethaib, S., Omar, R., Mustapa Kamal, S., & Radiah, D. (2015). Microwave-assisted pretreatment of lignocellulosic biomass: A review. *Journal of Engineering Science and Technology, 10*, 97–109.

Janusz, G., Pawlik, A., Sulej, J., Swiderska-Burek, U., Jarosz-Wilkolazka, A., & Paszczynski, A. (2017). Lignin degradation: Microorganisms, enzymes involved, genomes analysis and evolution. *FEMS Microbiology Reviews, 41*(6), 941–962. https://doi.org/10.1093/femsre/fux049

Jong, F. S. (2018). An overview of sago industry development, 1980s–2015. In H. Ehara, Y. Toyoda, & D. V. Johnson (Eds.), *Sago Palm: Multiple contributions to food security and sustainable livelihoods* (pp. 75–89). Springer Singapore.

Kapdan, I. K., & Kargi, F. (2006). Bio-hydrogen production from waste materials. *Enzyme and Microbial Technology, 38*(5), 569–582. https://doi.org/10.1016/j.enzmictec.2005.09.015
Kim, J. S., Lee, Y. Y., & Kim, T. H. (2016). A review on alkaline pretreatment technology for bioconversion of lignocellulosic biomass. *Bioresource Technology, 199*, 42–48. https://doi.org/10.1016/j.biortech.2015.08.085
Kumoro, A. C., Ngoh, G. C., Hasan, M., Ong, C. H., Teoh, E. C. (2008). Conversion of fibrous sago (*Metroxylon sagu*) waste into fermentable sugar via acid and enzymatic hydrolysis. *Asian Journal of Scientific Research*, 1. https://doi.org/10.3923/ajsr.2008.412.420
Amin, N. M., Sabli, N., Izhar, S., & Yoshida, H. (2019). A review: Sago Wastes and Its Applications. *Pertanika Journal of Science and Technology, 27*, 1841–1862.
McCay, M. H., Shafiee, S. (2020). 22—Hydrogen: An energy carrier. In T. M. Letcher (Ed.), *Future energy* (3rd ed., pp. 475–493). Elsevier.
Mishra, P., & Das, D. (2014). Biohydrogen production from Enterobacter cloacae IIT-BT 08 using distillery effluent. *International Journal of Hydrogen Energy, 39*(14), 7496–7507. https://doi.org/10.1016/j.ijhydene.2013.08.100
Mogili, N. V., Murugesan, N., Ayothiraman, S., Gautam, R., Deshavath, N. N., & Reddy, E. R. (2022). Chapter 6— Biohydrogen production from wastewater and organic solid wastes. In C. M. Hussain, S. Singh, & L. Goswami (Eds.), *Waste-to-energy approaches towards zero waste* (pp. 165–195). Elsevier.
Mohamad Puad, N. I., Abd Rahim, N. F., & Azmi, A. S. (2018). Acid pretreatment of sago wastewater for biohydrogen production. In *E3S web conference* (Vol. 34, p. 02012).
Osman, A. I., Mehta, N., Elgarahy, A. M., Hefny, M., Al-Hinai, A., AaH, A.-M., et al. (2022). Hydrogen production, storage, utilisation and environmental impacts: A review. *Environmental Chemistry Letters, 20*(1), 153–188. https://doi.org/10.1007/s10311-021-01322-8
Puad, N. I. M., Sulaiman, S., Azmi, A. S., Shamsudin, Y., & Mel, M. (2015). Preliminary study on biohydrogen production by *E. coli* from sago waste. *Journal of Engineering Science and Technology, 10*, 12–21.
Puad, N. I. M., Rashidi, U. M. M., Azmi, A. S., & Wahab, M. F. A. (2020). Microwave-assisted acid pretreatment of sago wastewater for fermentable sugar production. In: *IOP conference series: Materials science and engineering* (Vol. 778(1), p. 012051). https://doi.org/10.1088/1757-899x/778/1/012051
Qazi, U. (2022). Future of hydrogen as an alternative fuel for next-generation industrial applications; challenges and expected opportunities. *Energies, 15*, 4741. https://doi.org/10.3390/en15134741
Ramprakash, B., & Muthukumar, K. (2015). Comparative study on the performance of various pretreatment and hydrolysis methods for the production of biohydrogen using Enterobacter aerogenes RM 08 from rice mill wastewater. *International Journal of Hydrogen Energy, 40*(30), 9106–9112. https://doi.org/10.1016/j.ijhydene.2015.05.027
Soares, J. F., Confortin, T. C., Todero, I., Mayer, F. D., & Mazutti, M. A. (2020). Dark fermentative biohydrogen production from lignocellulosic biomass: Technological challenges and future prospects. *Renewable and Sustainable Energy Reviews, 117*, 109484. https://doi.org/10.1016/j.rser.2019.109484
Sunaryanto, R., Handayani, B., & Ratu, S. (2013). Enzymatic and acid hydrolysis of sago starch for preparation of ethanol production. *Microbiology Indonesia, 7*, 68–74. https://doi.org/10.5454/mi.7.2.4
Taherzadeh, M. J., & Karimi, K. (2008). Pretreatment of lignocellulosic wastes to improve ethanol and biogas production: A review. *International Journal of Molecular Sciences, 9*(9), 1621–1651. https://doi.org/10.3390/ijms9091621
Thangavelu, S. K., Ahmed, A. S., & Ani, F. N. (2014). Bioethanol production from sago pith waste using microwave hydrothermal hydrolysis accelerated by carbon dioxide. *Applied Energy, 128*, 277–283. https://doi.org/10.1016/j.apenergy.2014.04.076
Ulhiza, T. A., Puad, N. I. M., & Azmi, A. S. (2017). Review on biohydrogen production by dark fermentativebacteria using starch-containing waste as a substrate. *Journal of Advanced Research in Materials Science, 38*(1), 21–31.

Ulhiza, T. A., Mohamad Puad, N. I., & Azmi, A. S. (2018). Optimization of culture conditions for biohydrogen production from sago wastewater by Enterobacter aerogenes using Response Surface Methodology. *International Journal of Hydrogen Energy, 43*(49), 22148–22158. https://doi.org/10.1016/j.ijhydene.2018.10.057

Urbaniec, K., & Bakker, R. R. (2015). Biomass residues as raw material for dark hydrogen fermentation—a review. *International Journal of Hydrogen Energy, 40*(9), 3648–3658. https://doi.org/10.1016/j.ijhydene.2015.01.073

Vasco-Correa, J., Ge, X., & Li, Y. (2016). Chapter 24—biological pretreatment of lignocellulosic biomass. In S. I. Mussatto (Ed.), *Biomass fractionation technologies for a lignocellulosic feedstock based biorefinery* (pp. 561–585). Elsevier.

Woiciechowski, A. L., Nitsche, S., Pandey, A. P. A., & Soccol, C. R. (2002). Acid and enzymatic hydrolysis to recover reducing sugars from cassava bagasse: An economic study. *Brazilian Archives of Biology and Technology, 45*(3), 393–400. https://doi.org/10.1590/S1516-89132002000300018

Yunus, N., Jahim, J. M., Anuar, N., Abdullah, S. R. S., & Kofli, N. T. (2014). Batch fermentative hydrogen production utilising sago (Metroxylon sp.) starch processing effluent by enriched sago sludge consortia. *International Journal of Hydrogen Energy, 39*(35), 19937–19946. https://doi.org/10.1016/j.ijhydene.2014.10.015

Zhu, Z., Macquarrie, D. J., Simister, R., Gomez, L. D., & McQueen-Mason, S. J. (2015). Microwave assisted chemical pretreatment of miscanthus under different temperature regimes. *Sustainable Chemical Processes, 3*(1), 15. https://doi.org/10.1186/s40508-015-0041-6

Production of Zeolite as a Catalyst for Sustainable Aviation Fuel

Sarina Sulaiman, Muhammad Faris Zainol, Nurul Sakinah Engliman, Siti Hajar Binti Yusoc, Harumi Veny, and Siti Zubaidah

Abstract Sustainable aviation fuel or SAF is a non-conventional fuel for the aviation industry that is produced using different pathways such as Fisher-Tropsch (FT), Hydroprocessed Esters and Fatty Acids (HEFA), and Alcohol-to-Jet (ATJ). Four different methods are compared namely sol-gel, microwave synthesis, alkali Fusion and hydrothermal to produce zeolite for SAF production.

1 Introduction

Sustainable aviation fuel or SAF is a non-conventional fuel for the aviation industry that is produced using different pathways such as Fisher-Tropsch (FT), Hydroprocessed Esters and Fatty Acids (HEFA), and Alcohol-to-Jet (ATJ) (Goh et al. 2022). Sustainable Aviation Fuel (SAF) focuses on the efforts to mitigate carbon emissions and reduce their overall environmental impacts. The production of aviation fuel involves a reaction between a catalyst such as zeolites and a feedstock such as waste cooking oil.

Palm Kernel Shell (PKS) is a hard, fibrous material that is brown and is typically obtained from palm oil mills, where the palm kernel is separated from the shell using mechanical methods before the palm oil extraction from the fruit. According to Elham (2001), a single fruit of oil palms contains only around 27% of oil from its entirety, while 6–7% of the fruit composition is taken by PKS and the rest of the percentage is filled in by fiber and non-filled fruit bunch elements. On average, it is

S. Sulaiman (✉) · M. Faris Zainol · N. Sakinah Engliman · S. Hajar Binti Yusoc
Department of Chemical Engineering and Sustainability, Faculty of Engineering, International Islamic University Malaysia, Kuala Lumpur, Malaysia
e-mail: sarina@iium.edu.my

H. Veny
Faculty of Chemical Engineering, Universiti Teknologi MARA (UiTM), 40450 Shah Alam, Selangor, Malaysia

S. Zubaidah
Faculty of Chemical and Process Engineering Technology, Universiti Malaysia Pahang Al-Sultan Abdullah, Darul Makmur, Pahang, Malaysia

F. Adyani Ahmad Fuad et al. (eds.), *Circular Bioeconomy*,
https://doi.org/10.1007/978-981-97-7010-6_7

estimated that 4.7 million tons of PKS are generated as residues annually. PKS can be utilized to produce zeolites.

In zeolite production, it generally involves a combination of chemical synthesis, crystallization, post-synthesis treatment, and ion exchange. It is a highly demanding process as it consumes a significant amount of energy from its critical needs for high temperatures in every single unit operation (Parvulescu & Maurer 2022). Throughout recent years, zeolite has been constantly developed and relied upon as a catalyst for chemical reactions, mainly due to its attractive, inherent properties such as porous, crystalline structure, compositions, ion-exchange capability, and molecule selectivity that allow them to stay relevant in most industrial activity. Many methods of manufacturing and synthesizing zeolite have been explored and tested in the industry such as microwave heating. Microwave heating utilizes the function of a lab-scale microwave oven that is equipped with an electromagnetic radiation generator to radiate a wavelength between 0.1 and 1 m. In comparison with the conventional method, the microwave heating method has several advantages over traditional methods of zeolite synthesis, including faster reaction times, higher yields, and improved crystal size and purity, while also allowing for precise control of the reaction conditions (Zeng et al. 2021).

2 Zeolite

2.1 Background

Zeolites are a class of minerals that can be found in nature or produced synthetically. They possess a distinctive microporous architecture that renders them exceptionally valuable as adsorbents, catalysts, and ion exchangers. Parvulescu et al. (2022) have reported that the composition of zeolite comprises a tridimensional framework of silica and alumina tetrahedra, wherein the interstitial spaces are occupied by cations or water molecules (Parvulescu & Maurer 2022). Zeolites can selectively adsorb, or exchange ions or molecules based on their size, shape, and polarity due to their porous nature. This makes them useful in a wide range of applications from wastewater treatment to petrochemical refining, among others (Fig. 1). Zeolites can be produced from a variety of natural and synthetic raw materials, including volcanic ash, clays, and industrial waste streams. The utilization of palm kernel shells as a precursor in the production of zeolites is motivated by their cost-effectiveness and abundance in substantial quantities. Additionally, it presents an opportunity for efficient waste management.

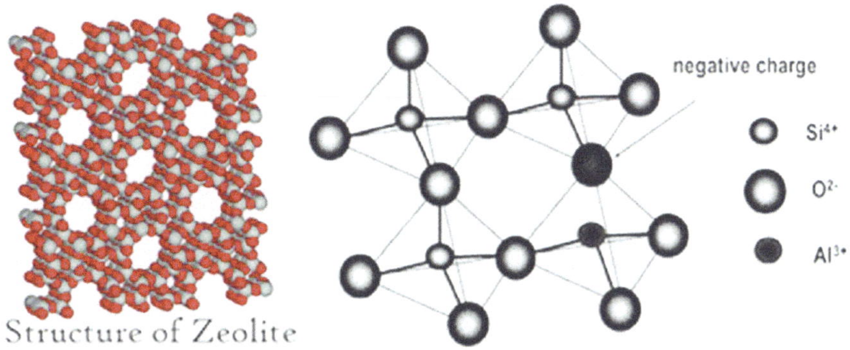

Fig. 1 General structure of zeolite

2.2 *Properties of Zeolite*

Zeolites exhibit several important properties due to their unique framework structure that varies based on the material and process condition employed (Khaleque et al. 2020). The property of zeolite is different between natural and synthetic, given that the synthetic one is produced in a lab within a short duration with the luxury of controlling the pore size, crystallinity, composition, and other properties to meet a specified purpose, while the natural one required longer synthesized time without the capability of altering the properties. The properties of synthetic zeolite of various types are summarized in Table 1.

These zeolites differ primarily in terms of aluminum concentration, which influences and affects the zeolite molecule's framework and ionic properties (Khaleque et al. 2020). Zeolite A is mainly possessed by a window structure of 8 rings with a channel diameter ranging from 0.29 to 0.38 nm (Feng et al. 2005). Furthermore, the cation exchange capacity is directly proportional to the alumina and sodium content of zeolite. The chemical composition of Zeolite consists of alumina, silica, and sodium with a variance in the cationic form that differentiates the unit formula for each type of zeolite.

3 Synthesis of Zeolite

Zeolite can be synthesized with very specific properties for various applications. When synthesizing zeolite, it can be made in various types of raw materials, as long as the materials have enough of the prerequisite elemental composition, such as carbon, oxygen, silica, and alumina (Khaleque et al. 2020) Although, the synthesized zeolite will fall under the category of synthetic zeolite since it is processed and enhanced by other auxiliary components to meet the desired properties. Several methods have been studied by previous researchers using different approaches as shown in Table 2.

Table 1 Properties of synthetic zeolites (Khaleque et al. 2020)

Framework	Cationic form	Typical unit cell formula	Window	Channel diameter (nm)	Cation exchange capacity (meq/g)	References
A	Na	Na12[(AlO_2)12(SiO_2)12]	8-ring (obstructed)	0.38	5.3	Bajpai and Tyagi (2007), Feng et al. (2005), Malekpour et al. (2008), Solache-Ríos (2001)
	Ca	Ca2Na2[(AlO_2)12(SiO_2)12]	8-ring (free)	0.44	–	
	K	K12[(AlO_2)12(SiO_2)12	8-ring (obstructed)	0.29	–	
P	Na	Na6(AlO_2)6(SiO_2)6 · 5 H_2O	GIS (gismondine)	ca. 0.3	–	Cardoso et al. (2015, Nissen et al. (2000)
X	Na	Na86[(AlO_2)86(SiO_2)106]	12-ring	0.84	4.5	Feng et al. (2005), Solache-Ríos (2001), Su and Lu (2012)
	Ca	Ca40Na6[(AlO_2)86(SiO_2)106	12-ring	0.8	–	
Y	Na	Na56[(AlO_2)56(SiO_2)136]	12-ring	0.8	3.26	Solache-Ríos (2001)
	K	K56(AlO_2)56(SiO_2)136]	12-ring	0.8	–	
ZSM-5	Na	Na3[(AlO_2)3(SiO_2)93]	10-ring	0.6	2.07	Solache-Ríos (2001)

Table 2 Comparison of different methods of zeolite synthesis (Mamudu et al. 2021)

Synthesis method	Raw materials	Auxiliary materials	Reaction condition	Type of zeolite	Advantages	Limitations	References
Hydrothermal	Coal fly ash	KOH, Deionized water	160 °C for 24 h	Zeolite K	High Reactivity, low pollution and energy consumption, good crystal quality	Expensive autoclave, corrosive slurries	Lakiss et al. (2020)
	Rice husk ash	H_2O, NaOH, Al_2O_3, tetra proylammonium hydroide	150 °C for 48 h	ZSM-5 Zeolite			Sari et al. (2015)
Alkali fusion	Coal fly ash	NaOH/Fly ash ratio of 1:3	550 °C for 24 h	Zeolite NaX	Use of low-grade raw materials and production of high-quality anhydrous zeolite	High energy consumption and costs	Fukasawa et al. (2017)
Alkaline hydrothermal	Paper sludge	HCl, NaOH	90 °C for 4/8 h	Zeolite A	Use of low-grade raw materials and production of high-quality anhydrous zeolite	High energy consumption and costs	Rojas-Valencia et al. (2014), Wajima et al. (2007)
	Palm Kernel Shell	NaOH	80 °C for 9 h	Zeolite Y			Mamudu et al. (2021)
Alkaline leaching	Fly ash	NaOH, $NaAlO_2$, water	90 °C for 8 h	A–X zeolite	Allows impure raw materials and which offers very efficient products	Multistep, expensive and time consuming	El-Naggar et al. (2008)
Sol–gel	Rice husk ash	HCl, H_2O, NaOH, KOH, $NaAlO_2$	90 °C for 24 h	Zeolite NaY	A versatile method, less energy consumption, no need special and expensive equipment	Cost of precursors, difficult to avoid residual porosity and OH groups	Wittayakun et al. (2008)
Microwave synthesis	Coal fly ash	H_2O, NaOH, $NaAlO_2$	75 °C for 1 h	Na-A zeolite	Faster reaction, selective heating, higher temperature	Difficulty in heat force control, water evaporation	Tanaka et al. (2008a 2008b)

Table 2 shows several methods that can be used to synthesize zeolite. Based on the study from Lakiss et al. (2014), the researcher employs the conventional hydrothermal method for the project with coal fly ash as the feedstock. There are only two auxiliary materials that have been used by the researchers, that is potassium hydroxide and deionized water with the reaction condition being 24-h contact time at 160 °C. By comparing the reaction condition with other methods such as microwave heating and alkaline hydrothermal, the hydrothermal method is advantageous and proves to have high energy consumption while holding a long reaction time, which simply translates to increased operation cost.

3.1 Hydrothermal

Similarly, the study by Sari et al. (2015) that employs the same hydrothermal method using rice husk ash was also found to have the reaction condition of the same properties, that is high temperature and long process time (Sari et al. 2015). The physical properties of zeolite synthesized using the hydrothermal method can be seen in Table 3, referring to a study by Borel et al. (2017) that produced Y-type Zeolite. According to the research, the Y-type zeolite is observed to have a 15–1 silica-alumina ratio and a BET surface area of 830 m^2/g, which is the highest compared to the other zeolites in Table 3.

3.2 Alkali Fusion

In comparison with the hydrothermal method, the alkali-fusion method uses way higher energy compared to the rest (Khaleque et al. 2020). The reason behind the excessively high temperature (550 °C) is believed to be due to the presence of alkali metal hydroxides or carbonates and the silica source that needs to be melted first so that better mixing and interaction between components can be achieved (Fukasawa et al. 2017). For alkaline hydrothermal, it is a hybrid of the previous two methods that involves the use of an alkaline solution at elevated temperatures and pressures (Khaleque et al. 2020). As observed from Table 3, researchers Mamudu et al. (2021) decided to use this method while utilizing PKS as the raw material to synthesize a Y-type zeolite for the fluid-catalyzed cracking unit (FCCU). In the study, the researchers set a heating time of 9 h at 80 C using a furnace as the reaction condition, and as a result, they managed to synthesize the Zeolite Y with a high silicate content to meet the desired properties. However, despite the temperature being relatively lower than the previous two methods, it still translates to high energy consumption considering the long operation time taken to synthesize the zeolite. In terms of physical properties of the zeolite, researcher Ojha et al. (2004) reported a range of 1.68–1.98 for silica-alumina ratio and 383 m^2/g for BET surface area with respect to the zeolite X

Table 3 Properties of synthetic zeolites produced by different approaches (Mamudu et al. 2021)

Synthesis method	Structure	SiO_2/Al_2O_3 ratio	Ion exchange capacity (mg/g)	BET surface area (m^2/g)	Crystal size	Crystallinity (%)	Microspore volume (cm^3/g)	References
Hydrothermal	Zeolite Y	15	–	830	–	–	0.33	Borel et al. (2017)
Alkali fusion	Zeolite X	1.68–1.98	–	383	5.69 μm	125.93	–	Ojha et al. (2004)
Alkaline leaching	A–X zeolite blend	2.1	4.624	593.64	-	96	–	El-Naggar et al. (2008)
Sol–gel	MCM-22	36	–	525	2.0 μm	100	–	Wu et al. (2009)
Microwave synthesis	Zeolite Y	–	–	645	6–19 μm	–	0.322	Le et al. (2019a 2019b)

synthesized from alkali-fusion method. Zeolite also possessed a high percentage of crystallinity at 125% and a crystal size of 5.69 μm.

3.3 Sol–Gel

Similarly, for the sol–gel method, the method takes place at 90 C for 24 h, according to the study by Wittayakun et al. (2008). The researchers used rice husk ash as the raw material to produce NaY-type zeolite. The process mainly involves the formation of a sol or colloidal suspension of precursor species, followed by gelation and subsequent heat treatment to form zeolite crystals (Khaleque et al. 2020)In terms of the properties of the zeolite synthesized through the sol–gel method, it recorded to have high crystallinity and silica-alumina ratio, according to a study by Wu et al. (2009), which is perfect for adsorption process for wastewater treatment.

3.4 Microwave Synthesis

Meanwhile, for the microwave method, the researchers Tanaka et al. (2008a 2008b) employ a reaction condition of 75 C heating temperature for 1 h with coal fly ash as the raw material. This study shows that the microwave method has the lowest energy consumption and the most cost-efficient as the temperature and process time is lowered significantly compared to the other method. However, in contrast to the study by Le et al. (2019a 2019b) the heating temperature used is higher (150 °C) compared to the previous researchers despite using the same method. This is believed to be due to the shorter process time set by the researcher that requires higher energy to meet the desired result within the duration to produce Zeolite Y. The raw material used by the researchers was undisclosed for us to draw a comparison and more concrete reasoning behind the increase in the heating temperature. In terms of the properties of the synthesized zeolite using the microwave method, researcher Le et al. (2019a 2019b) managed to produce a Y-type Zeolite with a BET surface area of 645 m^2/g and crystal size of 6–19 μm. The high BET surface area indicates that the synthesized zeolite possessed a large surface that is available for adsorption and catalytic reactions, which is perfect for utilization in wastewater treatment and petroleum refining. Unfortunately, other key properties such as crystallinity and silica-alumina ratio were not provided by the researcher to build a comparison with the synthesized zeolite from other methods.

References

Bajpai, D., & Tyagi, V. K. (2007). Synthesis of fatty imidazolines based on palm fatty acids and diethylenetriamine through microwave irradiation and their characterization. *Heterocyclic Communications, 13*(6). https://doi.org/10.1515/HC.2007.13.6.377.

Borel, M., Dodin, M., Daou, T. J., Bats, N., Harbuzaru, B., & Patarin, J. (2017). SDA-free hydrothermal synthesis of high-silica ultra-nanosized zeolite y. *Crystal Growth and Design, 17*(3). https://doi.org/10.1021/acs.cgd.6b01564.

Cardoso, A. M., Paprocki, A., Ferret, L. S., Azevedo, C. M. N., & Pires, M. (2015). Synthesis of zeolite Na-P1 under mild conditions using Brazilian coal fly ash and its application in wastewater treatment. *Fuel, 139*. https://doi.org/10.1016/j.fuel.2014.08.016.

Elham, P. (2001). The production of palm kernel shell charcoal by the continuous kiln method. In *Thesis*.

El-Naggar, M. R., El-Kamash, A. M., El-Dessouky, M. I., & Ghonaim, A. K. (2008). Two-step method for preparation of NaA-X zeolite blend from fly ash for removal of cesium ions. *Journal of Hazardous Materials, 154*(1–3). https://doi.org/10.1016/j.jhazmat.2007.10.115.

Feng, W., Mu-Sen, L., Yu-Peng, L., & Yong-Xin, Q. (2005). A simple sol-gel technique for preparing hydroxyapatite nanopowders. *Materials Letters, 59*(8–9). https://doi.org/10.1016/j.matlet.2004.08.041.

Fukasawa, T., Karisma, A. D., Shibata, D., Huang, A. N., & Fukui, K. (2017). Synthesis of zeolite from coal fly ash by microwave hydrothermal treatment with pulverization process. *Advanced Powder Technology, 28*(3). https://doi.org/10.1016/j.apt.2016.12.006.

Goh, B. H. H., Chong, C. T., Ong, H. C., Seljak, T., Katrašnik, T., Józsa, V., Ng, J. H., Tian, B., Karmarkar, S., & Ashokkumar, V. (2022). Recent advancements in catalytic conversion pathways for synthetic jet fuel produced from bioresources. In *Energy conversion and management* (Vol. 251). https://doi.org/10.1016/j.enconman.2021.114974.

Khaleque, A., Alam, M. M., Hoque, M., Mondal, S., Haider, J. B., Xu, B., Johir, M. A. H., Karmakar, A. K., Zhou, J. L., Ahmed, M. B., & Moni, M. A. (2020). Zeolite synthesis from low-cost materials and environmental applications: A review. *Environmental Advances, 2*, 100019. https://doi.org/10.1016/J.ENVADV.2020.100019.

Lakiss, L., Gilson, J. P., Valtchev, V., Mintova, S., Vicente, A., Vimont, A., Bedard, R., Abdo, S., & Bricker, J. (2020). Zeolites in a good shape: Catalyst forming by extrusion modifies their performances. *Microporous and Mesoporous Materials, 299*. https://doi.org/10.1016/j.micromeso.2020.110114.

Le, T., Wang, Q., Pan, B., Ravindra, A. V., Ju, S., & Peng, J. (2019a). Process regulation of microwave intensified synthesis of Y-type zeolite. *Microporous and Mesoporous Materials, 284*, 476–485. https://doi.org/10.1016/J.MICROMESO.2019.04.029.

Le, T., Wang, Q., Ravindra, A. V., Li, X., & Ju, S. (2019b). Microwave intensified synthesis of Zeolite-Y from spent FCC catalyst after acid activation. *Journal of Alloys and Compounds, 776*. https://doi.org/10.1016/j.jallcom.2018.10.316.

Malekpour, A., Millani, M. R., & Kheirkhah, M. (2008). Synthesis and characterization of a NaA zeolite membrane and its applications for desalination of radioactive solutions. *Desalination, 225*(1–3). https://doi.org/10.1016/j.desal.2007.02.096.

Mamudu, A., Emetere, M., Ishola, F., & Lawal, D. (2021). The production of zeolite y catalyst from palm kernel shell for fluid catalytic cracking unit. *International Journal of Chemical Engineering, 2021*. https://doi.org/10.1155/2021/8871228.

Nissen, L. R., Lepp, N. W., & Edwards, R. (2000). Synthetic zeolites as amendments for sewage sludge-based compost. *Chemosphere, 41*(1–2). https://doi.org/10.1016/S0045-6535(99)00420-8.

Ojha, K., Pradhan, N. C., & Samanta, A. N. (2004). Zeolite from fly ash: Synthesis and characterization. *Bulletin of Materials Science, 27*(6). https://doi.org/10.1007/BF02707285.

Parvulescu, A. N., & Maurer, S. (2022). Toward sustainability in zeolite manufacturing: An industry perspective. *Frontiers in Chemistry, 10*, 1516. https://doi.org/10.3389/FCHEM.2022.1050363/BIBTEX.

Rojas-Valencia, M.-N., Espejel-Ayala, F., Marin, S., & Ramírez-Zamora, R. M. (2014). Synthesis of zeolites using paper industry sludge as raw material. *Journal of Residual Science and Technology, 11*(1), 21–27.

Sari, Z. G. L. V., Younesi, H., & Kazemian, H. (2015). Synthesis of nanosized ZSM-5 zeolite using extracted silica from rice husk without adding any alumina source. *Applied Nanoscience (Switzerland), 5*(6). https://doi.org/10.1007/s13204-014-0370-x.

Solache-Ríos, M. (2001). Cation-exchange capacities of zeolites A, X, Y, ZSM-5 and Mexican erionite compared with the retention of cobalt and cadmium. *Journal of Radioanalytical and Nuclear Chemistry, 250*(1). https://doi.org/10.1023/A:1013280307656.

Su, F., & Lu, C. (2012). CO_2 capture from gas stream by zeolite 13X using a dual-column temperature/vacuum swing adsorption. *Energy and Environmental Science, 5*(10). https://doi.org/10.1039/c2ee22647b.

Tanaka, H., Fujii, A., Fujimoto, S., & Tanaka, Y. (2008a). Microwave-assisted two-step process for the synthesis of a single-phase Na-A zeolite from coal fly ash. *Advanced Powder Technology, 19*(1), 83–94. https://doi.org/10.1163/156855208X291783.

Tanaka, H., Fujimoto, S., Fujii, A., Hino, R., & Kawazoe, T. (2008). Microwave assisted two-step process for rapid synthesis of Na-A zeolite from coal fly ash. *Industrial and Engineering Chemistry Research, 47*(1). https://doi.org/10.1021/ie070603k.

Wajima, T., Ishimoto, H., Kuzawa, K., Ito, K., Tamada, O., Gunter, M. E., & Rakovan, J. F. (2007). Material conversion from paper-sludge ash in NaOH, KOH, and LiOH solutions. *American Mineralogist, 92*(7). https://doi.org/10.2138/am.2007.2251.

Wittayakun, J., Khemthong, P., & Prayoonpokarach, S. (2008). Synthesis and characterization of zeolite NaY from rice husk silica. *Korean Journal of Chemical Engineering, 25*(4). https://doi.org/10.1007/s11814-008-0142-y.

Wu, Y., Ren, X., & Wang, J. (2009). Facile synthesis and morphology control of zeolite MCM-22 via a two-step sol-gel route with tetraethyl orthosilicate as silica source. *Materials Chemistry and Physics, 113*(2–3). https://doi.org/10.1016/j.matchemphys.2008.08.008.

Zeng, X., Hu, X., Song, H., Xia, G., Shen, Z. Y., Yu, R., & Moskovits, M. (2021). Microwave synthesis of zeolites and their related applications. *Microporous and Mesoporous Materials, 323*, 111262. https://doi.org/10.1016/J.MICROMESO.2021.111262.

Environmental

Bioconversion of Palm Oil Mill Effluent (POME) into Bioethanol

Amanatuzzakiah Abdul Halim and Nazlina Haiza Mohd Yasin

Abstract The bioconversion of Palm Oil Mill Effluent (POME) into bioethanol has emerged as an environmentally friendly and economically viable solution for the valorization of this abundant waste stream. POME, a by-product of the palm oil industry, has been a major environmental concern due to its high organic content and potential to cause water pollution. However, recent advancements in biotechnology have enabled the conversion of POME into bioethanol, a renewable and sustainable biofuel with great economic potential. This study explores the economic feasibility of bioethanol production from POME and examines the bioconversion steps involved in the process. The first crucial step is the pretreatment of POME to enhance its biodegradability and reduce inhibitory compounds. Various pretreatment methods, such as physical, chemical and biological treatment, have shown promising results in improving the overall efficiency of POME conversion. Subsequently, hydrolysis is employed to break down complex carbohydrates present in POME into fermentable sugars and finally the fermentation process where the fermentable sugars are converted into ethanol by specific microorganisms. The unique characteristics of each method involved in the process are discussed, and the comparisons among the studies are critically reviewed. Additionally, the challenges and future prospects of POME conversion into bioethanol are explored. Ultimately, this waste-to-biofuel conversion holds the promise of addressing environmental concerns, curbing greenhouse gas emissions and improving the overall sustainability of the palm oil industry.

Keywords POME · Bioconversion · Bioethanol · Waste-to-biofuel

A. Abdul Halim (✉)
Department of Chemical Engineering and Sustainability, Faculty of Engineering, International Islamic University of Malaysia (IIUM), Jalan Gombak, Kuala Lumpur 53100, Malaysia
e-mail: amana@iium.edu.my

N. H. Mohd Yasin
Department of Biological Sciences and Biotechnology, Faculty of Science and Technology, Universiti Kebangsaan Malaysia, UKM Bangi, 43600 Bangi, Selangor, Malaysia
e-mail: nazlinayasin@ukm.edu.my

F. Adyani Ahmad Fuad et al. (eds.), *Circular Bioeconomy*,
https://doi.org/10.1007/978-981-97-7010-6_8

1 Introduction

Elaeis guineensis, commonly known as the oil palm tree, produces palm oil, the most widely used vegetable oil globally, utilized in both food and non-food industries (Bala et al. 2014). As one of the leading agricultural sectors in Malaysia, the palm oil industry produces an average of over 13 million tonnes of crude palm oil (CPO) annually. Malaysia is the second-largest exporter of palm oil after Indonesia, generating billion of dollars in revenue and creating jobs for the community (Zaidi et al. 2022). However, large-scale oil palm production results in substantial amount of palm oil mill effluent (POME), a dark brown viscous liquid rich in colloidal suspended matter and characterized by a strong, unpleasant odour. It has been demonstrated that approximately 2.5–3.75 tonnes of POME are produced for every tonne of CPO (Abdul et al. 2013). In 2022, Malaysia produced 18.5 million tonnes of CPO, generating around 68 million tonnes of POME (Malaysian Palm Oil Board 2022). Figure 1 shows a blackish POME sample in a raceway pond.

Raw POME is characterized as a viscous, acidic (pH 3.4–5.2), hot (80–90 °C), brownish/blackish colloidal mixture with an unpleasant odour (Mohammad et al. 2021). Table 1 presents the characteristics of POME obtained from different studies, along with the standard discharge limits set by the Malaysian Department of Environment under the Environmental Quality Regulations. POME has a high organic load (oil and grease: 130–18,000 mg/L, COD: 15,000–100,000 mg/L, BOD: 10,250–43,750 mg/L) but is nontoxic, consisting of 95-96% water and 2–4% suspended

Fig. 1 Blackish palm oil mill effluent (POME) sample in a raceway pond

solids (Bala et al. 2014; Ibrahim et al. 2015). The high organic content in POME poses significant environmental challenges. Therefore, the palm oil mill discharge standards, as outlined in Table 1, were developed by the Malaysian Department of Environment to control and reduce the pollution caused by the palm oil industry.

Several biological, chemical and physical methods have been reported for treating POME in industries. However, most biological methods, such as the ponding system, anaerobic digestion and phycoremediation (POME treatment using microalgae) (Samsudin et al. 2018, 2019) have been favored by industries due to their cost-effectiveness and technical feasibility. This article discusses the utilization of POME as a feedstock for bioethanol production. The processes required for the conversion of POME to bioethanol are explained in detail. Overall, using POME as a feedstock for value-added products helps sustain future revenue in industries and supports biofuel policies as a renewable energy source.

Table 1 Analysis of POME characteristics from different studies as compared to standard discharge limit by the Malaysian Department of Environment (DOE)

Parameter	Other Studies			Discharge standard by DOE (Kamyab et al. 2018)
	Mohammad et al. 2021	Kamyab et al. 2018	Najafpour et al. 2006	
Temperature (°C)	80–90	–	–	45
pH	3.4–5.2	4.7	4.7	5–9
Oil and grease (mg/L)	130–18,000	4000	4000	50
Biochemical Oxygen Demand (BOD) (mg/L)	10,250–43,750	25,000	25,000	100
Chemical Oxygen Demand (COD) (mg/L)	15,000–100,000	50,000	50,000	–
Total solids (mg/L)	–	40,500	40,500	–
Suspended solids (mg/L)	5000–54,000	18,000	18,000	400
Total volatile solid (mg/L)	–	–	34,000	–
Ammonia nitrogen (mg/L)	–	–	35	100
Total nitrogen (mg/L)	–	750	750	–

2 Bioethanol Production from POME

2.1 Introduction

Bioethanol is primarily produced using fermentable sugar from sugar crops such as sugarcane, sugar beet and sweet sorghum. This is the conventional method for producing bioethanol from sugar or starchy materials and is deemed as the first generation of bioethanol from renewable sources. Alternatively, bioethanol can be produced from non-food crops, including agricultural waste, wood chips and municipal wastes. This second-generation bioethanol is derived from lignocellulosic biomass, where cellulose is separated and converted into fermentable sugars. Utilizing resources from non-food crops allows for a more sustainable biofuel supply, addressing the limitations of first-generation bioethanol (Lee & Lavoie 2013).

In general, bioethanol is produced by fermenting sugars (mostly glucose) using yeast strains such as *Saccharomyces cerevisiae.* However, converting biomass materials into bioethanol is more complex. Lignocellulosic biomass, which consists of lignin, hemicellulose and cellulose, is intertwined with each other to form fibrous matrix (Fig. 2). This complex structure gives biomass its high strength. There are three main steps involved in producing bioethanol from biomass which are the pretreatment, hydrolysis and fermentation process. In the pretreatment step, lignocellulosic biomass is delignified to release cellulose and hemicellulose. This is followed by the hydrolysis of starchy materials into simple sugars, which are then fermented by microorganisms to produce bioethanol. Table 2 shows the chemical composition of various palm oil biomass.

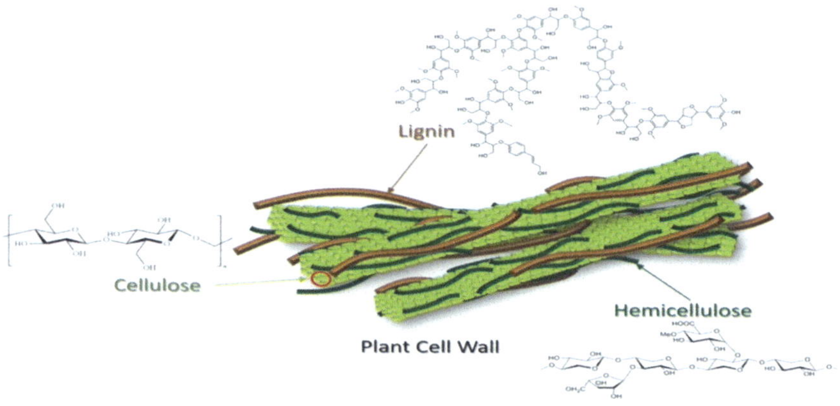

Fig. 2 Structure of lignocellulosic biomass (Jensen et al. 2017)

Table 2 Chemical composition of oil palm solid residues (Hamzah et al. 2019; Mun et al. 2008)

Type of biomass	Chemical components (% Dry Weight)				
	Cellulose	Hemicellulose	Lignin	Extractives	Ash
Empty fruit bunches	38.3	35.3	22.1	2.7	1.6
Mesocarp fibre	33.9	26.1	27.7	6.9	3.5
Palm kernel shell	20.8	22.7	50.7	4.8	1.0
Oil palm frond	30.4	40.4	21.7	1.7	5.8
Oil palm trunk	34.5	31.8	25.7	3.7	4.3
POME	39.6	23.3	25.0	–	–

2.2 *Economic Potential*

The global bioethanol production in 2023 is estimated at 112 billion litres (Renewable Fuels Association 2024), with an anticipated annual growth rate of 2% (Sharma et al. 2020). The main contributors are U.S.A (53%), Brazil (28%), EU (5%), China (3%) and Canada (2%). Even though 96% of the total world biofuels production in 2020 was dominated by the 1st generation biofuels, this statistic is more likely to change in the next decade as the 2nd generation bioethanol will increase its contribution into the global market through commercialization with growth in the conversion technology from inexpensive waste (Robak & Balcerek 2020).

Each step of the conversion process has to be optimized to ensure the cost of converting the biomass waste to the final product to be as low as possible to maintain profitability. Even though the cost of acquiring the waste is very minimal, the overall production cost is still high due to the nature of the recalcitrant materials used as the feedstock which requires intensive labour and high capital cost (Himmel et al. 2007). Eijck et al. (2014) conducted a study on economic performance of 2nd generation bioethanol using eucalyptus, switchgrass, poplar, wheat straw and rice straw. The ethanol conversion technology route is based on physical and acid pretreatment followed by enzyme saccharification and the fermentation process. The overall production cost is estimated to be at 16–26 $/GJ in 2020 and expected to be further reduced in the future with improved conversion technology and scaling up production facilities. Moreover the findings indicate that the key to produce economical bioethanol is through the optimization of the conversion process which accounted for 35–65% of total cost.

The potential bioethanol derived from POME can be estimated based on the calculation reported by Goh et al. (2010). The calculation was derived theoretically from the main recovery sugar (glucose and xylose) of the lignocellulosic material that provides the feedstock for the fermentation process. The content of bioethanol is calculated separately for each sugar extracted from cellulose and hemicellulose with their individual yield percentage, recovery efficiency and fermentation efficiency. All constants values used in the calculations are based on previous reported study

Table 3 Potential bioethanol production from POME feedstock in Malaysia from 2018–2022

	2018	2019	2020	2021	2022
Crude palm oil (tonnes)	19,516,141	19,858,367	19,140,613	18,116,354	18,453,420
POME (tonnes)	72,209,722	73,475,958	70,820,268	67,030,510	68,277,654
Bioethanol from cellulose (tonnes)	8,320,467	8,466,371	8,160,365	7,723,685	7,867,389
Bioethanol from hemicellulose (tonnes)	3,918,090	3,986,796	3,842,699	3,637,067	3,704,737
Total bioethanol (tonnes)	12,238,558	12,453,167	12,003,064	11,360,752	11,572,126

(Demirbaş 2005). The capacity of bioethanol production from POME can be calculated based on Eqs. (1) and (2). Data on the yearly production of crude palm oil were collected from MPOB (Malaysian Palm Oil Board 2022), whereby for every tonne of crude palm oil produced, 3.7 tonnes of POME were generated (Prasertsan & Prasertsan 1996). The potential production of bioethanol from POME based on crude palm oil production from 2018–2022 is summarized in Table 3.

$$\begin{aligned}&\text{Bioethanol yield from cellulose (tonnes)}\\&\quad=\text{cellulose (tonnes)}\times\text{theoretical yield }(0.5111)\\&\quad\times\text{glucose recovery efficiency }(0.76)\times\text{fermentation efficiency }(0.75)\end{aligned}\tag{1}$$

$$\begin{aligned}&\text{Bioethanol yield from hemicellose (tonnes)}\\&\quad=\text{hemicellose (tonnes)}\times\text{theoretical yield }(0.5175)\\&\quad\times\text{xylose recovery efficiency }(0.90)\times\text{fermentation recovery }(0.50)\end{aligned}\tag{2}$$

2.3 *Bioconversion Process of POME into Ethanol*

2.3.1 Pretreatment

POME has huge potential in bioethanol production, however high lignin contents in the organic material slow down the decomposition of cellulose and hemicellulose and limit hydrolytic enzymes efficiency. Generally, pretreatment step is required to break down the complex structure of lignin and hemicelluloses, and assists the conversion of crystalline cellulose into amorphous cellulose (Cardona & Sánchez 2007). These would increase the surface area and porosity of the materials thus improve the accessibility of the hydrolytic enzyme and simultaneously maximize sugar recovery. To achieve an efficient and economical pretreatment process, several guidelines can be followed as suggested by Taherzadeh and Karimi (2008) which are to minimize the formation of inhibitory compounds, avoid destruction of hemicellulose and cellulose, minimize energy used, reduce the cost of equipment and material, minimize chemicals usage and finally opt for compatible chemicals in the process. There are several methods for pretreatment process which can be categorized into physical-physio-chemical, chemical and biological methods. Table 4 summarizes various pretreatment methods for lignocellulosic material with their advantages and disadvantages information.

Physical Pretreatment

Physical pretreatment process involves mechanical process such as milling, grinding and chipping to reduce particle size, crystallinity index and polymerization degree (Vasić et al. 2021). Smaller particle size in the range of 10–30 and 0.2–2 mm can be achieved through chipping and milling/grinding respectively (Sun & Cheng 2002). However this process requires certain energy requirement based on the final particle size and biomass characteristics. Small particle size offers higher surface area which enhances the enzyme-cellulose affinity and therefore increase the rate of hydrolysis (Gu et al. 2018). Several studies reported the effect of biomass particle size on the pretreatment, hydrolysis as well as fermentation process (Lan et al. 2020; Zichao Wang et al. 2020; Li et al. 2020; Kapoor et al. 2019). The results obtained highlight the significance of particle size factor in these processes to achieve higher conversion at optimum cost.

Chemical Pretreatment

The common chemical pretreatment process is using either acid or base. Acid such as H_2SO_4 and HCl works by hydrolyzing hemicellulose and exposes the cellulose for hydrolysis process to be converted into sugar. It can also disrupt lignin but at a low efficiency rate (Wang et al. 2015). Concentrated acid treatment (70% acid content) produced high sugar recovery up to 90% of both hemicellulose and cellulose sugars compared to dilute acid hydrolysis (Shahbazi & Zhang 2010). However, this method is less desirable due to the toxic and corrosive characteristic of the acid which requires special corrosive resistant equipment. In addition to that, formation of inhibitors affects the subsequent process in bioethanol production and the need to develop acid recovery system is the reason why concentrated acid treatment is not recommended. Alternatively dilute acid can be used to hydrolyze hemicellulose into its monomer at high temperature to ensure the efficiency of the process (Keskin et al. 2019). Mahmod et al. (2017) reported higher yield of glucose from POME treated with 0.8% phosphoric acid and 1% nitric acid. Both dilute acid treatments resulted in higher biohydrogen production up to 97% compared to the untreated POME. Dilute HCl and H_2SO_4 were also used in other studies using POME for the production of biohydrogen (Seengenyoung 2013; Kamal et al. 2012; Azman et al. 2016). Some by-products of dilute acid treatment were detected such as furfural, hydroxymethylfurfural (HMF)

Table 4 Advantages and disadvantages of various pretreatment method for lignocellulosic material (Rizal et al. 2018)

Pretreatment category	Pretreatment method	Advantages	Disadvantages
Physical	Milling, grinding, chipping, shredding	Easily reduces the substrate size and increases the surface area	High energy requirement
		Short process duration	High cost for equipment and machinery setup
			Required additional pretreatment as it does not remove lignin
Chemical	Acid	Short process duration	Corrosive on the equipment
		Possible to fully hydrolyze the entire solid	Not economically feasible
	Base	Short process duration	Effective if the biomass has a low lignin content
		Selectively attacks the lignin	Less effective if the lignin content is high in biomass
			Usually requires high temperature to dissolve lignin
	Ionic liquids (ILs)	Inexpensive	Inefficient on the recovery of ILs
		Not toxic to enzymes and fermentation	
		Can be recovered and recycled	
Physico-chemical	Steam explosion	No chemicals used	Incomplete destruction of the lignin-carbohydrate matrix
		Low energy input	Risk of condensation and precipitation of soluble lignin components
		Environment-friendly	Destruction of a portion of xylan in hemicellulose
			Inhibitor formed at higher temperatures

(continued)

Table 4 (continued)

Pretreatment category	Pretreatment method	Advantages	Disadvantages
	Liquid hot water	Hydrolysis of hemicellulose and removal of lignin	Amount of solubilized product is higher
		Lower temperature used	
		Fewer inhibitors produced at high temperature	
		Effective for large-scale application	
	Superheated steam	Improved energy efficiency	Partial hemicelluloses degradation
		Low environmental impact when condensate is reused	Risk of condensation and precipitation of soluble lignin components
		Time saving	Not effective in removing lignin
		Environment-friendly	
		Cost-effective for large scale	
Biological	Fungi	Selectively degrades lignin	Slow process and therefore it needs long process duration
		Environment-friendly	Requires large space
		Requires mild operating conditions	Some carbohydrate fraction is consumed by the microorganism
			Needs to be conducted in sterile conditions
	Ligninolytic enzymes	Selectively degrades lignin and does not digest the carbohydrate structure	Enzyme cost is very high
		Environment-friendly	
		Requires mild operating conditions	
		Can be conducted in non-sterile conditions	

and aromatic lignin degradation compounds that affect the fermentation process, however, the generated compounds are minimal and it also depends on the process temperature used (Saha et al. 2005).

For base treatment, typically NaOH or KOH are used. The process enhances lignin and hemicellulose degradation while reducing the polymerization and crystallinity of cellulose (Bochmann 2019). The dissolution of lignin increases the internal surface area and subsequently provides a greater degree of access for enzyme digestion compared to acid treatment that dissolves only carbohydrates (Rizal et al. 2018). Superior effect of base treatment on POME was demonstrated by Seengenyoung (2013) in biohydrogen production. In this study, the yield of biohydrogen was three times greater with POME treated with 1.5% w/v NaOH compared to raw POME and doubled compared to POME pretreated with 1.5% w/v HCl. Similarly, higher yield of biohydrogen was obtained for POME pretreated with NaOH compared to H_2SO_4 (Kamal et al. 2012).

Other than acid and base treatment, ionic liquids (ILs) and deep eutectic solvents (DESs) can be used for biomass pretreatment process. These green solvents are an ideal alternatives as it offers several advantages which include inexpensive cost, requires low/mild temperature condition, nontoxic towards enzymes and cells and excellent in removing lignin/hemicellulose (Rizal et al. 2018; Lugani et al. 2020). ILs comprise of cations and anions components with strong electrostatic bonding. The ionic components can be altered to suit with the process and have excellent thermal stability with low volatility (Robak & Balcerek 2020). Many studies have demonstrated the efficiency of ILs in lignocellulosic material pretreatment process such as rice straw (Mohammadi et al. 2019), eucalyptus (Wang et al. 2018), oil palm frond (Financie et al. 2016) and switchgrass (Xia et al. 2014). DESs on the other hand are a mixture of organic compounds between various quaternary ammonium salts (as hydrogen acceptor) and a hydrogen donor such as carboxylic acids, amino acids, urea, amines or carbohydrates (Zhang et al. 2016). The utilization of these compounds has shown to improve high hydrolysis and fermentation yields (Thi & Lee 2019; Procentese et al. 2015; Nagoor Gunny et al. 2014; Loong et al. 2018). Despite of the advantages of these green solvents in pretreatment of lignocellulosic material, there was a concern pertaining the recovery process which requires extra step. The additional cost incurred may limit their application in the large-scale biorefineries. Nevertheless, there are potentials for ILs treatment to be economically viable by providing efficient means to extract sugars from lignin with the integration of solvents recovery and recycling process (Konda et al. 2014).

Physico-Chemical Treatment

Physico-chemical treatment combines both physical and chemical interactions in the process. Among the common methods is steam explosion whereby the method utilizes high pressure and temperature range. Several studies using oil palm biomass were reported. The range of pressure used was 20–50 bars with temperature range of 160–240 °C for 1–120 min (Rizal et al. 2018). These conditions allow the steam to penetrate into the porous structure of the fibrous biomass and create explosion due to the decompression process. This treatment helps in the degradation of lignocellulosic materials which improves the condition for the subsequent hydrolysis process. Alternatively liquid hot water pretreatment can be applied. This method operates in liquid phase at a high temperature range of 160–250 °C and low pressure setting around 5 bar (Taylor et al. 2019). This process results in the degradation of the majority of hemicellulose with no usage of chemicals and no toxic inhibitors formation (Mosier et al. 2005; Lynd et al. 2002; Banerjee et al. 2010).

Biological Treatment

Biological pretreatment of lignocellulosic biomass is an alternative approach that offers various advantages which include low capital costs, low energy requirement, environmental friendly and is performed under mild conditions (Rizal et al. 2018). This method uses certain microorganisms (fungi and bacteria) to reduce the recalcitrance of the biomass which improves its digestibility by hydrolytic enzymes. It can be performed through microorganisms that secrete ligninolytic enzymes or by using the enzymes directly. Most commonly used microbes are white-rot fungi, brown-rot fungi, soft-rot fungi and bacteria. Among these microbes, white-rot fungi is superior due to its capacity for lignin degradation and their unique ability to extensively mineralize lignin to carbon dioxide and water (Kirk & Farrell 1987). Several studies were conducted for fungal pretreatment of lignocellulosic biomass using *Phanerochaete chrysosporium, Pleurotus ostreatus, Ceriporiopsis subvermispora, Irpex lacteus and Trametes versicolor* (Vasco-Correa et al. 2016). Different species of fungi secrete different types of enzymes and have unique strategy of degradation (simultaneous rot or selective delignification) which affect the results of the treatment process (Reid 1995). Fungal treatment for 21 days of wheat straw using *Poria subvermispora and Irpex lacteus* reached 69 and 66% of glucose conversion from the cellulose available in the wheat straw, respectively, with an ethanol yield of 62% in both cases (Salvachúa et al. 2011). In another study of wheat straw, the pretreatment using *Phanerochaete chrysosporium* in submerged culture using inorganic salts and tween 80 reported highest lignin loss (25%) and approximately 250% higher efficiency for the total sugar release through enzymatic hydrolysis (Zeng et al. 2011). Considering the time required for this process is relatively long, combining different pretreatment and/or adding mediators or other chemicals is another option to enhance ligninolytic activity which improve the yield of the fungal pretreatment as well as to make it economically feasible (Vasco-Correa et al. 2016).

Applying enzyme extracts directly in the biomass for the depolymerization process of lignin can reduce the pretreatment time while simplifying the process compared to the microbial treatment. However there are concerns on the difficulties associated with heterogenous catalysis and complexity of the ligninolytic enzyme system which commonly requires a combination of several enzymes and the presence of mediators to ensure an efficient lignin degradation process (Vasco-Correa et al. 2016). Laccase enzyme has shown promising results for lignin degradation process in various biomass feedstock (Rico et al. 2014; Chen et al. 2010; Heap et al. 2014). Around 40% removal of lignin was observed in a study conducted using laccase pretreatment with a combination of superheated steam and size reduction process (hammer mill) in oil palm empty fruit bunch (OPEFB) and oil palm mesocorp fibre (OPMF) with a maximum generation of glucose yields of 72% which represents a 4.6 fold increase as compared to the untreated samples (Rizal et al. 2018). Similar work on OPEFB feedstock treated with laccase enzyme reported 58% lignin degradation with 161% increase in saccharification process compared with the untreated OPEFB (Rashid et al. 2019). Apart from laccase, other enzymes that involve in lignin degradation include manganese peroxidase (MnP) and lignin peroxidase (LiP) which have proven to increase the fermentable sugars yield (Ji et al. 2014).

In conclusion, many studies incorporated more than one pretreatment methods to optimize the sugar recovery for the fermentation process (Financie et al. 2016; Rusli et al. 2019). The pretreatment methods discussed here were not all been applied for POME which can be taken into consideration for future study. However, the choice of pretreatment should be based on several factors such as energy and time consumption, cost for chemicals and enzymes, capital cost for equipment, inhibitors generation and environmental effect of the waste produced.

2.3.2 Hydrolysis

Hydrolysis plays a crucial role in bioethanol production, as it releases the sugars needed for fermentation, which are then converted into ethanol in the final stage of the process. Hemicellulose and cellulose recovered from the pretreatment process can be broken down into simple sugars by enzymatic hydrolysis or chemically (acid and base treatment). Hydrolysis of hemicellulose releases mostly pentose sugars (eg: xylose, arabinose and rhamnose) and some hexose sugars (eg: mannose and galactose). On the other hand, hydrolysis of cellulose produces glucose. Due to the nature of cellulose that is highly resistant to hydrolysis, pretreatment of biomass is required to ensure an efficient hydrolysis process. Such pretreatment improves the accessibility of the hydrolytic enzymes and simultaneously enhances the recovery of fermentable sugars.

Cellulases enzymes (endoglucanase, cellobiohydrolase and β-glucosidase) are used to convert cellulose into glucose under mild conditions, such as temperature from 40–50 C and pH of 4.5–5 (Vasić et al. 2021). Enzyme hydrolysis of POME using a combination of cellulases (endoglucanase, CTec and xylanase) for 24 h produced maximum of 30.5 g/L of fermentable sugars (Nurul-Adela et al. 2016). The hydrolysis process was further optimized using response surface methodology (RSM) with optimum sugar production of 35.65 g/L obtained at 200 rpm agitation, 1% (v/v) enzyme dosage and 100% (v/v) substrate concentration (Bukhari et al. 2020). Commercial xylanase was utilized for POME hydrolysis performed at 50 °C for 12 h for the production of biogas. The highest concentration of glucose and xylose recorded were 11.98 and 10.89 g/L respectively which were 5.8 and 17 folds higher compared to the raw POME (Prasertsan et al. 2017). A threefold higher of reducing sugar was obtained at the loading of 2.4 FPU/mL cellulase enzyme extracted from Trichoderma reesei for POME hydrolysis as compared to the untreated POME conducted at 28 °C and pH 4 (Tajul Islam et al. 2017). POME hydrolysate containing a total sugar concentration of 15.40 g/L was obtained from enzymatic hydrolysis using 15 Unit/g TVS (Total Volatile Solid) at 50 °C for 18 h incubation with the hydrolysis yield of 0.35 g total sugars/g TVS (Nutongkaew et al. 2020). The hydrolysis enzyme was isolated from fungi *Trichoderma koningiopsis* consisting of CMCase and xylanase enzyme. The results indicated that enzyme hydrolysis treatment and co-digestion of POME hydrolysate with OPT enhances the methane yield from anaerobic fermentation of POME. Enhancement of fermentable sugar recovery can be done by formulating better enzyme mixtures and optimizing hydrolysis parameters. The effect of various combinations of enzyme mixtures on POME hydrolysis using commercial enzyme at different hydrolysis process parameters were studied by Silvamany group (Silvamany et al. 2015). The highest fermentable sugar yield of 34.4 g/L was produced with ratio enzymes of 0.33:0.33:0.33 (cellulase: cellobiose: Viscozyme-L), at enzyme loading of 3%, pH 4.8 and 48 h of incubation at 50 °C.

Despite of the efficient hydrolysis process, enzymatic hydrolysis is still considered expensive. Nearly 50% of the total production cost for lignocellulosic bioethanol is attributed to the enzymes, making it approximately 50% more expensive to produce compared to corn-starch-based ethanol, with an estimated production cost of US$0.26/L (Karimi & Chisti 2017). Enzyme recycling is one of the alternatives to shift the overall production process to be more economically feasible. Among the methods for enzyme recycling are sedimentation followed by ultra filtration or micro centrifugation, cation exchange chromatography, re-adsorption and immobilization (Binod et al. 2011). Centrifugation method on enzyme recovery shows favourable results whereby the majority of the initial enzyme activity was recovered in the initial stages of enzyme recycling (Moniruzzaman et al. 1997). However gradual decrease in enzyme activity was observed at the later stages which could be due to the thermal or mechanical inactivation. Enzyme immobilization method is very common to be used as an option for enzyme recycling. The reusability capacity of β-galactosidase immobilized on methacrylamide polymer carrier (Eupergit C) for lignocellulose hydrolysis was studied (Tu et al. 2006). The enzyme activity was maintained and relatively stable throughout six successive rounds of lignocellulosic hydrolysis. The stability of the immobilized enzyme was also improved at 65 °C as compared to the free enzyme while the profile of activity versus pH was unchanged.

In addition to that, enzyme hydrolysis is a slow process and is inhibited by the products of the hydrolysis such as glucose and cellobiose. Study on the inhibition effects of glucose and other sugar monomers in cellulase and β-galactosidase hydrolysis on softwood substrate were conducted by Xiao et al. (2004). As the glucose concentration increased in the softwood hydrolysate, dramatic increase was observed in the degree of inhibition on both β-galactosidase and cellulase activities. Addition of mannose, xylose and galactose during cellobiose hydrolysis did not show any inhibitory effects towards β-galactosidase but significant inhibitory effects on cellulase activity. This study suggested constant high-substrate supplementation of hemicellulose would minimize the end-product inhibition effect while obtaining high sugar recovery. Alternatively, enzyme hydrolysis process can be combined together with the fermentation process or also known as simultaneous saccharification and fermentation (SSF). Through this route, the product inhibitory effect in enzyme hydrolysis can be minimized with direct consumption of sugars during fermentation. Besides that, the implementation of SSF simplifies the need of separate vessel for the two processes which in turn reduces the capital investment. However, this method poses several drawbacks such as low bioethanol productivity due to the difficulties to achieve the optimum conditions for simultaneous enzyme hydrolysis and fermentation process and accumulation of ethanol also reported to severely inhibit the hydrolytic enzymes (Taherzadeh & Karimi 2007).

2.3.3 Fermentation

In the fermentation step, microorganisms use the available sugars produced during pretreatment and hydrolysis and convert it into bioethanol. These two processes are designed to optimize the fermentation process. Generally, yeast such as *S. cerevisiae* is used in alcohol production mostly in wine and brewery industries for hundreds of years. In addition to that, microorganisms such as bacteria (*Zymomonas mobilis Thermoanaerobacterium saccharolyticum, Thermoanaerobacter mathranii, Thermoanaerobacter ethanolicus*) and fungi (*Mucor indicus*) are other potential producers of bioethanol (Karimi & Chisti 2017). During fermentation, S. cerevisiae converts hexose sugars mainly glucose under anaerobic condition with ideal temperature and optimal pH of 30 °C and 4 respectively (Limayem & Ricke 2012). This conventional strain can generate high yield of ethanol (12.0–17.0% w/v; 90% of the theoretical) from hexose sugars (Bayrock & Ingledew 2001). Despite of high tolerance characteristic of *S. cerevisiae* in producing a high yield of ethanol, it is not able to ferment pentose sugars such as xylose which constitute a large proportion of sugar generated from lignocellulosic material. Similarly, *Zymomonas mobilis* bacterium can only ferment hexose sugar. The fermentation follows Entner-Doudoroff (ED) pathway which enables *Zymomonas mobilis* to produce less biomass and (~5%) more ethanol compared to *S. cerevisiae* (Wirawan et al. 2020). Alternatively, pentose fermenting yeasts include *Candida shehatae, Scheffersomyces stipitisa and Pachysolen tannophilus* (Robak & Balcerek 2020). Due to the limitation of fermenting certain sugars, various microorganisms have been metabolically engineered to improve performance in bioethanol production (Rogers et al. 2007; Den Haan et al. 2013; Nevoigt 2008). The co-culture system can also be employed by combining both hexose and pentose metabolizing strains in the fermentation (Chen 2011).

Statistical optimization study was conducted on bioethanol production from POME (Alam et al. 2009). Improved bioethanol yield was achieved at 51.3 g/L using single-step fermentation supplemented with 2% of wheat flour using mixed culture of *T. harzianum and S. cerevisiae* under optimum conditions of 30% pO_2, 32 °C and pH 6. The chemical oxygen demand (COD) was reduced to 89% after 4 days fermentation which reflected the biodegradation of POME through the consumption of soluble and insoluble organic substances by the microorganisms for ethanol production and cell growth. On the other hand, low ethanol yield was reported from POME fermentation using bacterial mixed cultures sourced from bovine rumen (Anggamulia et al. 2020). Approximately 10 g/L of ethanol was obtained from 5L fermentation run at pH 5 for 12 hours using anaerobic circular batch reactor. Besides that, spontaneous fermentation using different supplement was performed by Wakil et al. (2013) using direct POME sample for 21 days fermentation. Mixed cultures isolated from the fermentation were identified as *Bacillus carotarum, Bacillus lentus, Bacillus stearothermophilus, Bacillus pumilis, Micrococcus luteus, Yarrowia lipolytica, Candida intermedia, Candida tropicalis, Clavispora lusitaneae, Saccharomyces cerevisiae, Aspergillus flavus, Aspergillus niger, Aspergillus ochraceous and Penicillium* species. The highest ethanol yield recorded for unsupplemented fermentation was 3.95 g/L while the ethanol content increased to 18.17 g/L with fermentation supplemented with 30 g/L of glucose and 30 g/L of sugar cane bagasse for 12 days at pH of 8.5. In another study, *Clostridium acetobutylicum* grown in 90% sedimented POME recorded a production of Acetone-Butanol-Ethanol (ABE) of 4 g/L with an initial pH of 5.8 (Kalil et al. 2003). Although *C. acetobutylicum* is able to grow in liquid POME and produces ABE, the utilization of sedimented POME is preferable due to the higher content of lignocellulose and other insoluble materials which supported the growth and improves the production of the ABE. All in all, the ethanol yield reported by the studies varied based on the starting materials and operating conditions used during the fermentation. All studies reported here are using single-step bioconversion or also known as simultaneous saccharification and fermentation (SSF) whereby the process combines enzymatic hydrolysis with the fermentation process. The final yield of the ethanol is dependent on the sugar content available during the fermentation which highly correlated with the microorganisms present that secrete certain enzymes for the hydrolysis process. The optimum operating conditions for ethanol fermentation are estimated to be within a temperature range of 30–35 °C and a pH range of 5–6.

3 Challenges and Future Prospects

The 7th Sustainable Development Goals (SDG) by the United Nations embarks on developing affordable, sustainable and clean energy sources. Bioethanol production from lignocellulosic materials in POME is a second-generation biofuel in compliance with the 7th SDG (Ocreto et al. 2021). However, there are a few challenges for the production and utilization of bioethanol as a biofuel source. The first challenge is the remote location of the plantation, making on-site ethanol production necessary to overcome logistical barriers for the POME biorefinery. Furthermore, the limited space in the plantations, where the ponding system has fully utilized all available area for POME treatment, restricts the possibility of setting up additional facilities for bioethanol production. The second challenge is that most palm oil industries prefer to utilize POME directly in anaerobic digesters to produce biogas which holds higher economic value as boiler fuel (Chin et al. 2013). Unlike bioethanol production, which requires pretreatment, the anaerobic digestion of POME for biogas production is a straightforward process. This was also supported by Entry Points Projects No. 5 (EPP5) under the Malaysia Palm Oil National Key Economic Area (NKEA), which urges the installation of biogas facilities to mitigate greenhouse gase emissions from palm oil industries (Loh et al. 2017). The third challenge is that bioethanol production from POME requires further optimization studies before it can be widely implemented in industries. While the organic content in POME has been fully utilized for bioethanol production, the handling and time-intensive processes at the pilot and industrial scales still demand additional refinement and optimization (Mohammad et al. 2021).

Despite the challenges mentioned, POME is a reliable biomass that can serve as a feedstock for bioenergy production. Utilizing POME for bioethanol production offers sustainable benefits, including socio-political advantages (such as job creation and community engagement) and environmental benefits (Hassan et al. 2019). Moreover, unlike corn and sugarcane, bioethanol from POME does not conflict with food security concerns (Awogbemi & Kallon 2022). Overall, the production of bioethanol from POME has significant potential for commercialization due to its advantages, market stability, and sustainability.

4 Conclusion

In conclusion, the production of ethanol from Palm Oil Mill Effluent (POME) offers significant environmental and economic benefits. By converting this waste stream into a valuable resource, the process helps reduce pollution and mitigates the negative impacts associated with POME disposal. Additionally, ethanol production from POME contributes to the sustainability of the palm oil industry by diversifying revenue streams and minimizing its environmental footprint.

References

Abdul, P. M., Jahim, J. M., Harun, S., Markom, M., Hassan, O., Mohammad, A. W., & Asis, A. J. (2013). Biohydrogen production from pentose-rich oil palm empty fruit bunch molasses: A first trial. *International Journal of Hydrogen Energy, 38*(35), 15693–15699. https://doi.org/10.1016/j.ijhydene.2013.05.050.

Alam, M. Z., Kabbashi, N. A., Nahdatul, S., & Hussin, I. S. (2009). Production of bioethanol by direct bioconversion of oil-palm industrial effluent in a stirred-tank bioreactor. *Journal of Industrial Microbiology and Biotechnology, 36*(6). https://doi.org/10.1007/s10295-009-0554-7.

Anggamulia, M. I., Syafila, M., Handajani, M., & Gumilar, A. (2020). The potential bio-conversion of palm oil mill effluent (POME) as bioethanol by steady-state anaerobic processes. *E3S Web of Conferences, 148*. https://doi.org/10.1051/e3sconf/202014802001

Awogbemi, O., & Kallon, D. V. V. (2022). Valorization of agricultural wastes for biofuel applications. *Heliyon*. https://doi.org/10.1016/j.heliyon.2022.e11117.

Azman, N. F., Abdeshahian, P., Al-Shorgani, N. K. N., Hamid, A. A., & Kalil, M. S. (2016). Production of hydrogen energy from dilute acid-hydrolyzed palm oil mill effluent in dark fermentation using an empirical model. *International Journal of Hydrogen Energy, 41*(37), 16373–16384. https://doi.org/10.1016/j.ijhydene.2016.05.085.

Bala, J. D., Lalung, J., & Ismail, N. (2014). Palm Oil Mill Effluent (POME) treatment microbial communities in an anaerobic digester: A review. *International Journal of Scientific and Research Publications, 4*(6), 1–24.

Banerjee, S., Mudliar, S., Sen, R., Giri, B., Satpute, D., Chakrabarti, T., & Pandey, R. A. (2010). Commercializing lignocellulosic bioethanol: Technology bottlenecks and possible remedies. *Biofuels, Bioproducts and Biorefining*. https://doi.org/10.1002/bbb.188.

Bayrock, D., & Ingledew, W. M. (2001). Changes in Steady State on Introduction of a Lactobacillus Contaminant to a Continuous Culture Ethanol Fermentation. *Journal of Industrial Microbiology and Biotechnology,* 27(1). https://doi.org/10.1038/sj.jim.7000159.

Binod, P., Janu, K. U., Sindhu, R., & Pandey, A. (2011). Hydrolysis of lignocellulosic biomass for bioethanol production. *Biofuels*. https://doi.org/10.1016/B978-0-12-385099-7.00010-3.

Bochmann, G. (2019). Storage and feedstock preparation. In *Substitute natural gas from waste: Technical assessment and industrial applications of biochemical and thermochemical processes*. https://doi.org/10.1016/B978-0-12-815554-7.00004-0.

Bukhari, N. A., Loh, S. K., Nasrin, A. B., & Jahim, J. M. (2020). Enzymatic hydrolysate of palm oil mill effluent as potential substrate for bioflocculant BM-8 production. *Waste and Biomass Valorization, 11* (1). https://doi.org/10.1007/s12649-018-0421-8.

Cardona, C. A., & Sánchez, Ó. J. (2007). Fuel ethanol production: Process design trends and integration opportunities. *Bioresource Technology*. https://doi.org/10.1016/j.biortech.2007.01.002.

Chen, S., Zhang, X., Singh, D., Yu., Hongbo, & Yang, X. (2010). Biological pretreatment of lignocellulosics: Potential, progress and challenges. *Biofuels*. https://doi.org/10.4155/bfs.09.13.

Chen, Y. (2011). Development and application of co-culture for ethanol production by co-fermentation of glucose and xylose: A systematic review. *Journal of Industrial Microbiology and Biotechnology*. https://doi.org/10.1007/s10295-010-0894-3.

Chin, M. J., Poh, P. E., Tey, B. T., Chan, E. S., & Chin, K. L. (2013). Biogas from Palm Oil Mill Effluent (POME): opportunities and challenges from Malaysia's perspective. *Renewable and Sustainable Energy Reviews, 26*, 717–726. https://doi.org/10.1016/j.rser.2013.06.008.

Demirbaş, A. (2005). Bioethanol from cellulosic materials: A renewable motor fuel from biomass. *Energy Sources, 27*(4). https://doi.org/10.1080/00908310390266643.

van Eijck, J., Batidzirai, B., & Faaij, A. (2014). Current and future economic performance of first and second generation biofuels in developing countries. *Applied Energy, 135*. https://doi.org/10.1016/j.apenergy.2014.08.015.

Financie, R., Moniruzzaman, M., & Uemura, Y. (2016). Enhanced enzymatic delignification of oil palm biomass with ionic liquid pretreatment. *Biochemical Engineering Journal, 110*. https://doi.org/10.1016/j.bej.2016.02.008.

Goh, C. S., Tan, K. T., Lee, K. T., & Bhatia, S. (2010). Bio-ethanol from lignocellulose: status, perspectives and challenges in Malaysia. *Bioresource Technology, 101*(13). https://doi.org/10.1016/j.biortech.2009.08.080.

Gu, H., An, R., & Bao, J. (2018). Pretreatment refining leads to constant particle size distribution of lignocellulose biomass in enzymatic hydrolysis. *Chemical Engineering Journal, 352*. https://doi.org/10.1016/j.cej.2018.06.145.

Den Haan, R., Kroukamp, H., Mert, M., Bloom, M., Görgens, J. F., & Van Zyl, W. H. (2013). Engineering saccharomyces cerevisiae for next generation ethanol production. *Journal of Chemical Technology and Biotechnology*. https://doi.org/10.1002/jctb.4068.

Hamzah, N., Tokimatsu, K., & Yoshikawa, K. (2019). Solid fuel from oil palm biomass residues and municipal solid waste by hydrothermal treatment for electrical power generation in Malaysia: a review. *Sustainability (Switzerland)*. https://doi.org/10.3390/su11041060.

Hassan, M. A., Farid, M. A. A., Shirai, Y., Ariffin, H., Othman, M. R., Samsudin, M. H., & Hasan, M. Y. (2019). Oil palm biomass biorefinery for sustainable production of renewable materials. *Biotechnology Journal, 14*(6). https://doi.org/10.1002/biot.201800394.

Heap, L., Green, A., Brown, D., Van Dongen, B., & Turner, N. (2014). Role of laccase as an enzymatic pretreatment method to improve lignocellulosic saccharification. *Catalysis Science and Technology, 4*(8). https://doi.org/10.1039/c4cy00046c.

Himmel, M. E., Ding, S. Y., Johnson, D. K., Adney, W. S., Nimlos, M. R., Brady, J. W., & Foust, T. D. (2007). Biomass recalcitrance: Engineering plants and enzymes for biofuels production. *Science*. https://doi.org/10.1126/science.1137016.

Ibrahim, R. I., Wong, Z. H., & Mohammad, A. W. (2015). Optimization and performance evaluation for nutrient removal from palm oil mill effluent wastewater using microalgae. *IOP Conference Series: Materials Science and Engineering, 78*, 12006. https://doi.org/10.1088/1757-899X/78/1/012006.

Jensen, C. U., Guerrero, J. K. R., Karatzos, S., Olofsson, G., & Iversen, S. B. (2017). Fundamentals of hydrofaction^{TM}: Renewable crude oil from woody biomass. *Biomass Conversion and Biorefinery, 7*(4). https://doi.org/10.1007/s13399-017-0248-8.

Ji, L., Yang, J., Fan, H., Yang, Y., Li, B., Yu, X., Zhu, N., & Yuan, H. (2014). Synergy of crude enzyme cocktail from cold-adapted cladosporium cladosporioides Ch2–2 with commercial xylanase achieving high sugars yield at low cost. *Biotechnology for Biofuels, 7*(1). https://doi.org/10.1186/preaccept-8898871331282242.

Kalil, M. S., Kit, P. W., Yusof, W. M. W., Sadazo, Y., & Rahman, R. A. (2003). Direct fermentation of palm oil mill effluent to acetone-butanol-ethanol by solvent producing clostridia. *Pakistan Journal of Biological Sciences, 6*(14). https://doi.org/10.3923/pjbs.2003.1273.1275.

S. A., Jahim, J. M., Anuar, N., Hassan, O., Daud, W. R. W., Mansor, M. F. M., & Rashid, S. S. (2012). Pre-Treatment Effect of Palm Oil Mill Effluent (POME) during Hydrogen production by a local isolate clostridium butyricum. *International Journal on Advanced Science, Engineering and Information Technology, 2*(4). https://doi.org/10.18517/ijaseit.2.4.214.

Kamyab, H., Chelliapan, S., Din, M. F. M, Rezania, S., Khademi, T., & Kumar, A. (2018). Palm Oil Mill Effluent as an environmental pollutant. In *Palm oil*. InTech. https://doi.org/10.5772/intechopen.75811.

Kapoor, M., Semwal, S., Satlewal, A., Christopher, J., Gupta, R. P., Kumar, R., Puri, S. K., & Ramakumar, S. S. V. (2019). The impact of particle size of cellulosic residue and solid loadings on enzymatic hydrolysis with a mass balance. *Fuel, 245*. https://doi.org/10.1016/j.fuel.2019.02.094.

Karimi, K., Chisti, Y. (2017). Bioethanol production and technologies. In *Encyclopedia of sustainable technologies*. https://doi.org/10.1016/B978-0-12-409548-9.10111-3.

Keskin, T., Abubackar, H. N., Arslan, K., & Azbar, N. (2019). Biohydrogen production from solid wastes. *Biohydrogen*. https://doi.org/10.1016/b978-0-444-64203-5.00012-5

Kirk, T. K., & Farrell, R. L. (1987). Enzymatic 'combustion': The microbial degradation of lignin. *Annual Review of Microbiology*. https://doi.org/10.1146/annurev.mi.41.100187.002341

Konda, N. M., Shi, J., Singh, S., Blanch, H. W., Simmons, B. A., & Klein-Marcuschamer, D. (2014). Understanding cost drivers and economic potential of two variants of ionic liquid pretreatment for cellulosic biofuel production. *Biotechnology for Biofuels, 7*(1). https://doi.org/10.1186/1754-6834-7-86.

Lan, T. Q., Wang, S. R., Li, H., Qin, Y. Y., & Yue, G. J. (2020). Effect of lignin isolated from p-toluenesulfonic acid pretreatment liquid of sugarcane bagasse on enzymatic hydrolysis of cellulose and cellulase adsorption. *Industrial Crops and Products, 155*. https://doi.org/10.1016/j.indcrop.2020.112768

Lee, R. A., & Lavoie, J. M. (2013). From first- to third-generation biofuels: challenges of producing a commodity from a biomass of increasing complexity. *Animal Frontiers, 3*(2). https://doi.org/10.2527/af.2013-0010.

Li, X., Xiong, N., Wang, X., Dai, X., Guo, Y., & Dong, B. (2020). New insight into volatile sulfur compounds conversion in anaerobic digestion of excess sludge: Influence of free ammonia nitrogen and thermal hydrolysis pretreatment. *Journal of Cleaner Production, 277*. https://doi.org/10.1016/j.jclepro.2020.123366.

Limayem, A., & Ricke, S. C. (2012). Lignocellulosic biomass for bioethanol production: Current perspectives, potential issues and future prospects. *Progress in Energy and Combustion Science*. https://doi.org/10.1016/j.pecs.2012.03.002.

Loh, S. K., Nasrin, A. B., Mohamad Azri, S., Nurul Adela, B., Muzzammil, N., Daryl Jay, T., & Stasha, E. (2017). Biogas capture—A means of reducing greenhouse gas emissions from palm oil mill effluent. *Oil Palm Bulletin*, 27–36.

Loong, L. Y., Wu., Ta Yeong, Yang, G. H., Ang, L. Y., New, E. K., Siow, L. F., Md., Jamaliah, Jahim, A. W., & Mohammad, and Wen Hui Teoh. (2018). Deep eutectic solvent and inorganic salt pretreatment of lignocellulosic biomass for improving xylose recovery. *Bioresource Technology, 249*. https://doi.org/10.1016/j.biortech.2017.07.165.

Lugani, Y., Rai, R., Prabhu, A. A., Maan, P., Hans, M., Kumar, V., Kumar, S., Chandel, A. K., & Sengar, R. S. (2020). Recent advances in bioethanol production from lignocelluloses: A comprehensive review with a focus on enzyme engineering and designer biocatalysts. *Biofuel Research Journal, 7*(4). https://doi.org/10.18331/BRJ2020.7.4.5.

Lynd, L. R., Weimer, P. J., van Zyl, W. H., & Pretorius, I. S. (2002). Microbial cellulose utilization: fundamentals and biotechnology. *Microbiology and Molecular Biology Reviews, 66*(4). https://doi.org/10.1128/mmbr.66.4.739.2002.

Mahmod, S. S., Jahim, J. M., & Abdul, P. M. (2017). Pretreatment conditions of Palm Oil Mill Effluent (POME) for thermophilic biohydrogen production by mixed culture. *International Journal of Hydrogen Energy, 42*(45). https://doi.org/10.1016/j.ijhydene.2017.07.178.

Malaysian Palm Oil Board. (2022). Overview of the Malaysian Palm Oil Industry. https://bepi.mpob.gov.my.

Mohammad, S., Baidurah, S., Kobayashi, T., Ismail, N., & Leh, C. P. (2021). Palm Oil Mill effluent treatment processes—A Review. *Processes, 9*(5), 739. https://doi.org/10.3390/pr9050739.

Mohammadi, M., Shafiei, M., Karimi, K., Abdolmaleki, A., Mikkola, J. P., & Larsson, C. (2019). Improvement of ethanol production from birch and spruce pretreated with 1-H-3-methylmorpholinium chloride. *Electronic Journal of Biotechnology, 41*. https://doi.org/10.1016/j.ejbt.2019.07.004.

Moniruzzaman, M., Dale, B. E., Hespell, R. B., & Bothast, R. J. (1997). Enzymatic hydrolysis of high-moisture corn fiber pretreated by AFEX and recovery and recycling of the enzyme complex. *Applied Biochemistry and Biotechnology—Part A Enzyme Engineering and Biotechnology, 67*(1–2). https://doi.org/10.1007/BF02787846.

Mosier, N., Wyman, C., Dale, B., Elander, R., Lee, Y. Y., Holtzapple, M., & Ladisch, M. (2005). Features of promising technologies for pretreatment of lignocellulosic biomass. *Bioresource Technology, 96*(6). https://doi.org/10.1016/j.biortech.2004.06.025.

Mun, W. K., Rahman, N. A., Abd-Aziz, S., Sabaratnam, V., & Hassan, M. A. (2008). Enzymatic hydrolysis of palm oil mill effluent solid using mixed cellulases from locally isolated fungi. *Research Journal of Microbiology, 3*(6). https://doi.org/10.3923/jm.2008.474.481.

Nagoor Gunny, A. A., Arbain, D., Mohamed Daud, M. Z., & Jamal, P. (2014). Synergistic action of deep eutectic solvents and cellulases for lignocellulosic biomass hydrolysis. *Materials Research Innovations, 18*. https://doi.org/10.1179/1432891714Z.000000000933.

Najafpour, G. D., Zinatizadeh, A. A. L., Mohamed, A. R., Hasnain Isa, M., & Nasrollahzadeh, H. (2006). High-rate anaerobic digestion of palm oil mill effluent in an upflow anaerobic sludge-fixed film bioreactor. *Process Biochemistry, 41*(2), 370–379. https://doi.org/10.1016/j.procbio.2005.06.031.

Nevoigt, E. (2008). Progress in metabolic engineering of saccharomyces cerevisiae. *Microbiology and Molecular Biology Reviews, 72*(3). https://doi.org/10.1128/mmbr.00025-07.

Nurul-Adela, B., Nasrin, A. B., & Loh, S. K. (2016). Palm oil mill effluent as a low-cost substrate for bioflocculant production by bacillus marisflavi NA8. *Bioresources and Bioprocessing, 3*(1). https://doi.org/10.1186/s40643-016-0096-6.

Nutongkaew, T., Prasertsan, P., O-Thong, S., Chanthong, S., & Suyotha, W. (2020). Improved methane production using lignocellulolytic enzymes from trichoderma koningiopsis TM3 through co-digestion of palm oil mill effluent and oil palm trunk residues. *Waste and Biomass Valorization, 11*(10). https://doi.org/10.1007/s12649-019-00838-z.

Ocreto, J. B., Chen, W.-H., Ubando, A. T., Park, Y.-K., Sharma, A. K., Ashokkumar, V., Ok, Y. S., Kwon, E. E., Rollon, A. P., De Mark Daniel, G., & Luna. (2021). A Critical review on second- and third-generation bioethanol production using microwaved-assisted heating (MAH) pretreatment. *Renewable and Sustainable Energy Reviews, 152*, 111679. https://doi.org/10.1016/j.rser.2021.111679.

Prasertsan, P., Khangkhachit, W., Duangsuwan, W., Mamimin, C., & O-Thong, S. (2017). Direct hydrolysis of palm oil mill effluent by xylanase enzyme to enhance biogas production using two-steps thermophilic fermentation under non-sterile condition. *International Journal of Hydrogen Energy, 42*(45). https://doi.org/10.1016/j.ijhydene.2017.05.140.

Prasertsan, S., & Prasertsan, P. (1996). Biomass residues from palm oil mills in Thailand: An overview on quantity and potential usage. *Biomass and Bioenergy, 11*(5). https://doi.org/10.1016/S0961-9534(96)00034-7.

Procentese, A., Johnson, E., Orr, V., Campanile, A. G., Wood, J. A., Marzocchella, A., & Rehmann, L. (2015). Deep eutectic solvent pretreatment and subsequent saccharification of corncob. *Bioresource Technology, 192*. https://doi.org/10.1016/j.biortech.2015.05.053.

Rashid, S. S., Huan, H. S., Rahim, M. H. B. A., Sikder, M. B. H., Ahmed, A. H., & Sasi. (2019). Laccase enzyme immobilization on ferromagnetic support material for oil palm empty fruit bunches (OPEFB) pretreatment. *IOP Conference Series: Materials Science and Engineering, 627*. https://doi.org/10.1088/1757-899X/627/1/012025.

Reid, I. D. (1995). Biodegradation of lignin. *Canadian Journal of Botany, 73* (S1). https://doi.org/10.1139/b95-351.

Renewable Fuels Association. (2024). 2024 Ethanol Industry Outlook. https://ethanolrfa.org/resources/annual-industry-outlook

Rico, A., Rencoret, J., Del Río, J. C., Martínez, A. T., & Gutiérrez, A. (2014). Pretreatment with laccase and a phenolic mediator degrades lignin and enhances saccharification of eucalyptus feedstock. *Biotechnology for Biofuels, 7*(1). https://doi.org/10.1186/1754-6834-7-6.

Rizal, N. F. A. A., Ibrahim, M. F., Zakaria, M. R., Abd-Aziz, S., Yee, P. L., & Hassan, M. A. (2018). Pre-treatment of oil palm biomass for fermentable sugars production. *Molecules (Basel, Switzerland)*. https://doi.org/10.3390/molecules23061381.

Rizal, N. F. A. A., Ibrahim, M. F., Zakaria, M. R., Bahrin, E. K., Abd-Aziz, S., & Hassan, M. A. (2018). Combination of superheated steam with laccase pretreatment together with size reduction to enhance enzymatic hydrolysis of oil palm biomass. *Molecules, 23*(4). https://doi.org/10.3390/molecules23040811.

Robak, K., & Balcerek, M. (2020). Current state-of-the-art in ethanol production from lignocellulosic feedstocks. *Microbiological Research*. https://doi.org/10.1016/j.micres.2020.126534.

Rogers, P. L., Jeon, Y. J., Lee, K. J., & Lawford, H. G. (2007). Zymomonas mobilis for fuel ethanol and higher value products. *Advances in Biochemical Engineering/Biotechnology, 108*. https://doi.org/10.1007/10_2007_060.

Rusli, N. D., Azmi, M. A., Mat, K., Hasnita, C. H., Wan-Zahari, M., Azhar, K., Zamri-Saad, M., & Hassim, H. A. (2019). The effect of physical and biological pre-treatments of oil palm fronds on in vitro ruminal degradability. *Pertanika Journal of Tropical Agricultural Science, 42*(2).

Saha, B. C., Iten, L. B., Cotta, M. A., & Victor Wu, Y. (2005). Dilute acid pretreatment, enzymatic saccharification and fermentation of wheat straw to ethanol. *Process Biochemistry, 40*(12). https://doi.org/10.1016/j.procbio.2005.04.006.

Salvachúa, D., Prieto, A., López-Abelairas, M., Lu-Chau, T., Martínez, A. T., & Martínez, M. J. (2011). Fungal pretreatment: An alternative in second-generation ethanol from wheat straw. *Bioresource Technology, 102*(16). https://doi.org/10.1016/j.biortech.2011.05.027.

Samsudin, A., Azmi, A. S., Nawi, M. N. M., & Halim, A. A. (2018). Wastewater treatment by microalgae-bacteria co-culture system. *Malaysian Journal of Microbiology, 14*(3).

Samsudin, A., Azmi, A. S., Nawi, M. N. M., & Halim, A. A. (2019). Nutrients and chemical oxygen demand (COD) removals by microalgae -bacteria co-culture system in Palm Oil Mill Effluent (POME). *IIUM Engineering Journal, 20*(2). https://doi.org/10.31436/iiumej.v20i2.1109.

Seengenyoung, J. (2013). Biohydrogen production from palm oil mill effluent pretreated by chemical methods using thermoanaerobacterium-rich sludge. *Iranica Journal of Energy and Environment, 4*(4). https://doi.org/10.5829/idosi.ijee.2013.04.04.01.

Shahbazi, A., & Zhang, B. (2010). Dilute and concentrated acid hydrolysis of lignocellulosic biomass. In *Bioalcohol production: Biochemical conversion of lignocellulosic biomass.* https://doi.org/10.1533/9781845699611.2.143.

Sharma, B., Larroche, C., & Dussap, C. G. (2020). Comprehensive assessment of 2G bioethanol production. *Bioresource Technology*. https://doi.org/10.1016/j.biortech.2020.123630.

Silvamany, H., Harun, S., Mumtaz, T., & Jahim, J. M. (2015). Recovery of fermentable sugars from palm oil mill effluent via enzymatic hydrolysis. *Jurnal Teknologi, 77*(33). https://doi.org/10.11113/jt.v77.7016.

Sun, Y., & Cheng, J. (2002). Hydrolysis of lignocellulosic materials for ethanol production: A review. *Bioresource Technology, 83*(1). https://doi.org/10.1016/S0960-8524(01)00212-7.

Taherzadeh, M. J., & Karimi, K. (2007). Enzyme-based hydrolysis processes for ethanol from lignocellulosic materials: A review. *BioResources*. https://doi.org/10.15376/biores.2.4.707-738.

Taherzadeh, M. J., & Karimi, K. (2008). Pretreatment of lignocellulosic wastes to improve ethanol and biogas production: A review. *International Journal of Molecular Sciences*. https://doi.org/10.3390/ijms9091621.

Islam, M. S. T., Alam, M. Z., Al-Mamun, A., Elgharbawy, A. A., & Riyadi, F. A. (2017). Development of enzymatic pretreatment of palm oil mill effluent for monomers towards biogas production. *Journal of Advanced Research in Materials Science, 38.*

Taylor, M. J., Alabdrabalameer, H. A., & Skoulou, V. (2019). Choosing physical, physicochemical and chemical methods of pre-treating lignocellulosic wastes to repurpose into solid fuels. *Sustainability (Switzerland)*. https://doi.org/10.3390/su11133604.

Thi, S., & Lee, K. M. (2019). Comparison of Deep Eutectic Solvents (DES) on pretreatment of oil palm empty fruit bunch (OPEFB): Cellulose digestibility, structural and morphology changes. *Bioresource Technology, 282.* https://doi.org/10.1016/j.biortech.2019.03.065.

Tu, M., Zhang, X., Kurabi, A., Gilkes, N., Mabee, W., & Saddler, J. (2006). Immobilization of β-glucosidase on Eupergit C for lignocellulose hydrolysis. *Biotechnology Letters, 28*(3). https://doi.org/10.1007/s10529-005-5328-3.

Vasco-Correa, J., Ge, X., & Li, Y. (2016). Biological pretreatment of lignocellulosic biomass. In *Biomass fractionation technologies for a lignocellulosic feedstock based biorefinery*. https://doi.org/10.1016/B978-0-12-802323-5.00024-4.

Vasić, K., Knez, Z., & Leitgeb, M. (2021). Bioethanol production by enzymatic hydrolysis from different lignocellulosic sources. *Molecules, 26*(3). https://doi.org/10.3390/molecules26030753.

Wakil, S., Fasiku, S., Adelabu, A., & Onilude, A. (2013). Production of bioethanol from spontaneous fermentation of Palm Oil Mill Effluent (POME). *Researcher, 5.*

Wang, Z., Shao, S., Zhang, C., Lu, D., Ma, H., & Ren, X. (2015). Pretreatment of vinegar residue and anaerobic sludge for enhanced hydrogen and methane production in the two-stage anaerobic system. *International Journal of Hydrogen Energy, 40*(13). https://doi.org/10.1016/j.ijhydene.2015.02.029.

Wang, Z., Hou, X., Sun, J., Li, M., Chen, Z., & Gao, Z. (2018). Comparison of ultrasound-assisted ionic liquid and alkaline pretreatment of eucalyptus for enhancing enzymatic saccharification. *Bioresour Technol, 254.* https://doi.org/10.1016/j.biortech.2018.01.021.

Wang, Z., He, X., Yan, L., Wang, J., Hu., Xiaolong, Qi., Sun, & Zhang, H. (2020). Enhancing enzymatic hydrolysis of corn stover by twin-screw extrusion pretreatment. *Indus Crops Prod, 143.* https://doi.org/10.1016/j.indcrop.2019.111960.

Wirawan, F., Cheng, C. L., Lo, Y. C., Chen, C. Y., Chang, J. S., Leu, S. Y., & Lee, D. J. (2020). Continuous cellulosic bioethanol co-fermentation by immobilized Zymomonas mobilis and suspended Pichia stipitis in a two-stage process. *Appl Energy, 266*. https://doi.org/10.1016/j.apenergy.2020.114871.

Xia, S, Baker, G. A., Li, H., Ravula, S., & Zhao, H. (2014). Aqueous ionic liquids and deep eutectic solvents for cellulosic biomass pretreatment and saccharification. *RSC Advances, 4*(21). https://doi.org/10.1039/c3ra46149a.

Xiao, Z., Zhang, X., Gregg, D. J., & Saddler, J. N. (2004). Effects of sugar inhibition on cellulases and β-glucosidase during enzymatic hydrolysis of softwood substrates. In *Applied biochemistry and biotechnology—Part A enzyme engineering and biotechnology* (Vol. 115). https://doi.org/10.1385/ABAB:115:1-3:1115..

Zaidi, M. A. S., Karim, Z. A., & Zaidon, N. A. (2022). External and internal shocks and the movement of palm oil price: SVAR evidence from Malaysia. *Economies, 10*(1). https://doi.org/10.3390/economies10010007.

Zeng, J., Singh, D., & Chen, S. (2011). Biological pretreatment of wheat straw by phanerochaete chrysosporium supplemented with inorganic salts. *Bioresource Technology, 102*(3). https://doi.org/10.1016/j.biortech.2010.11.008.

Zhang, C. W., Xia, S. Q., & Ma, P. S. (2016). Facile pretreatment of lignocellulosic biomass using deep eutectic solvents. *Bioresour Technol, 219*. https://doi.org/10.1016/j.biortech.2016.07.026

Agro-waste Lignocellulose Valorization for Sustainable Circular Bioeconomy

Ainil Hawa Jasni, Azlin Suhaida Azmi, Usame Alasali, Noor Illi Mohamad Puad, Fathilah Ali, and Yusilawati Ahmad Nor

Abstract The use of agro-waste in industrial manufacturing has grown significantly due to overpopulation and the progressive bioeconomy model. Because of the wide range of waste applications and abundant resources, many businesses have begun large-scale production. This study reviewed the various agro-wastes and their applications. There have been several reports of agro-waste being utilized as value-added bioeconomy crops. In this paper, coconut fibre (CF) in the potential characteristic for the manufacture of lactic acid (LA) has also been studied. More specifically, current research and development in utilizing bio-agro-waste via the specific chemical treatment of ethylenediamine (EDA) were also discussed to improve the potential fermentability of the materials as substrate in LA production. As a result of the research, CF pretreatment using EDA modified the chemical structure with significant lignin and hemicellulose removal further supporting this claim. As a result, the surface characteristic of EDA-pre-treated CF was more accessible to hydrolytic enzymes, improving fermentation results. This paper addresses the importance of lignocellulose as a potential sustainable raw material for producing LA via EDA pretreatment for fermentation use. This discovery is useful in supporting the bioeconomy's agro-waste valorization process.

Keywords Agrowaste · Bioeconomy · Cellulose · Coconut fibre · Lactic acid and valorization

1 Introduction

Biotechnological valorization of agriculture or industrial wastes for chemical production is an innovative technique to reduce the costing and contribute to a circular bioeconomy. A circular bioeconomy is a holistic approach to economic development that benefits enterprises, society, and the environment, all at the same time.

A. H. Jasni · A. S. Azmi (✉) · U. Alasali · N. I. Mohamad Puad · F. Ali · Y. Ahmad Nor
Department of Chemical Engineering and Sustainability, Kulliyyah of Engineering, International Islamic University of Malaysia (IIUM), Jalan Gombak, 53100 Kuala Lumpur, Malaysia
e-mail: azlinsu76@iium.edu.my

F. Adyani Ahmad Fuad et al. (eds.), *Circular Bioeconomy*,
https://doi.org/10.1007/978-981-97-7010-6_9

A circular economy, in contrast to the 'take-make-waste' linear paradigm, is regenerative by design and seeks to eventually divorce growth from the consumption of scarce resources (BGB 2021). However, a circular bioeconomy is fueled by natural resources. It's a new economic model that stresses the utilization of renewable natural capital and concentrates on waste minimization, as opposed to the large range of non-renewable, fossil-based products now in use (The Center for International Forestry Research (CIFOR) and World Agroforestry (ICRAF) 2021). After defining what an economy is, this study route delves into the details of the circular economy concept, such as the distinction between biological and technical materials, the various options for keeping resources and products in use, and the history of the concept. It is, nevertheless, at risk of developing in limited, unsustainable ways. Diverse expert and stakeholder input, alternative solutions, and noneconomic value are all necessary for a sustainable road to a circular bioeconomy. Finally, the advantages of transitioning from a linear to a circular economy are emphasized (Reike et al., 2018). Shifting the energy and extractive industries to circular and resource-efficient models can result in significant cost savings.

For instance, in Nairobi, Kenya, 9 out of 10 homes cook on charcoal. In rural areas, producing charcoal to serve the city is a vital source of income (Akolgo 2021). In bags and furnaces, fine dust is frequently left behind. This dust had been collected and combined with a starch-binding agent to create briquettes for use as fuel. The selling of these briquettes, which were made from what was once deemed waste, has the potential to generate revenue (The Center for International Forestry Research (CIFOR) and World Agroforestry (ICRAF) 2021). Meanwhile in Malaysia, palm oil empty fruit bunch (EFB), as a low-cost biomass resource, has been utilized as an essential feedstock and composting material (Abdulrazik et al. 2017). Other than that, wood-based products are replacing carbon-intensive, non-renewable materials in a sustainable circular bioeconomy concept. For example, the building industry, which is dominated by concrete and steel, may become more environmentally friendly by including wood. Wood-based fibres could replace synthetic fibres in the textile sector, cutting carbon emissions. Figure 1 is the circular bioeconomy of the well-being cycle.

2 Agro-waste

This section focuses on agro-waste production by Malaysians. Based on the Department of Statistics Malaysia (2020) in Fig. 2, rice is the highest crop per capita consumption (PCC) followed by coconut, durian, banana, pineapple, and others. The higher the PCC, the higher the waste produced by each of these crops. Based on the statistics, this chapter will look at what kind of agricultural waste is generated from the selected crop/vegetable and the utilization or product from the waste (DOSM, 2023). Rice is the largest consumption in Malaysia, followed by Coconut and Durian. Rice is the staple crop of the country; thus, it is ranked as no.1 and the production is being done locally. Meanwhile, Table 1 summarizes the selected crop

Fig. 1 The circular bioeconomy cycle

and its respective value-added product which can be derived from the waste which supports the circular bioeconomy.

3 Valorization of Lignocellulose Biomass Resource

In other cases, many palm oil mills in Malaysia utilize palm Bioenergy as the simplest or the quickest value-added product which can be derived from biomass agricultural or agro-waste. The major example of bioenergy is listed in Fig. 3. The product

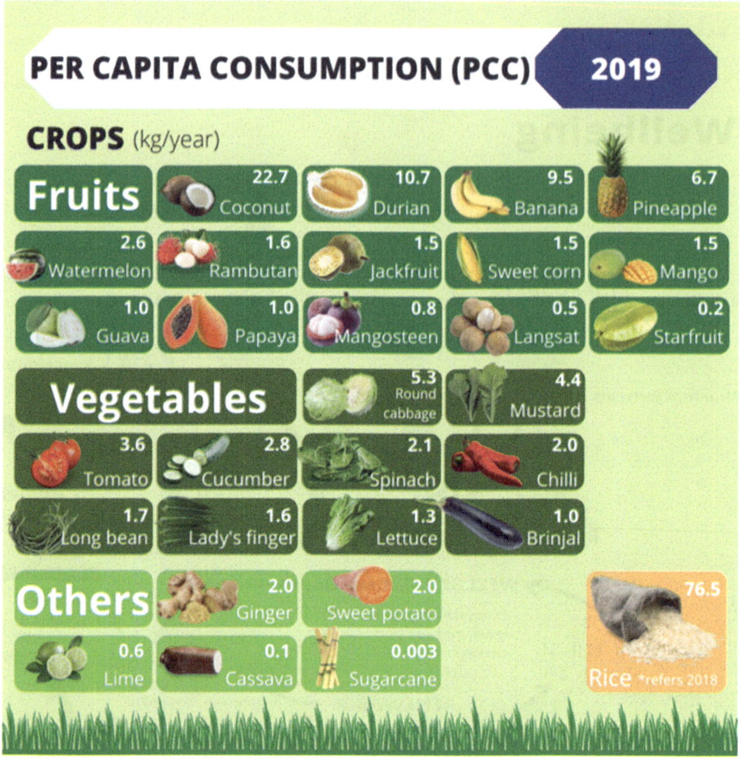

Fig. 2 PCC of agricultural fruits and vegetables in Malaysia in 2019

Table 1 Added value crops use in the bioeconomy

Crop	Waste	Product	References
Rice	Husk	Cement replacement and green concrete	[8]
		Rice husk briquettes	[9]
	Straw	Epoxy-cellulose nanocomposite for military application	[10]
Coconut	Shell	Coconut shell charcoal	[11]
		Briquettes	[11]
	Husk	Husk biochar	[12]
Durian	Skin	Activated carbon	[13]
Banana	Leave	Thermoplastic composite	[14]
	Stem	Cellulose for military application	[15]
	Peel	Biochar	

varies from biomass resource fuel for fuel switching to bio-compressed natural gas (BioCNG) (Malode et al. 2021). For instance, the production of polyester as an ideal strategy for agro-waste disposal had been implemented by SIRIM Malaysia Berhad. The SIRIM pilot plant produces 2,000L of polyhydroxyalkanoates (PHA) materials from palm oil mill wastewater and crude palm kernel oil (Teysha 2018) to burn boilers for power generation. Another example is using rice as a biomass resource for power generation. Perlis, a state in Malaysia, generates 10 MW of power from rice husks through a gasification process that meets the requirement for 30,000 households (Zareei et al. 2017). Based on Fig. 3, bioenergy is the lowest value economy value in the bioeconomy map, followed by bio-agriculture, eco-product such as bioplastics, and biochemicals are the highest values. However, valorization of the biomass to biochemicals takes a longer conversion and is much more complicated compared to bioenergy. Instead, managing these wastes, agriculture, forests, wetlands, and bioenergy sustainably and comprehensively could help to mitigate more than 30% of global climate change while also addressing pressing health issues (The Center for International Forestry Research (CIFOR) and World Agroforestry (ICRAF) 2021). Indeed, since the world's focus has shifted towards sustainable development and renewable resources. Nowadays, the technological readiness to convert wastes into value-added products is developing very fast as it demonstrates significant marketing potential as well as economic feasibility.

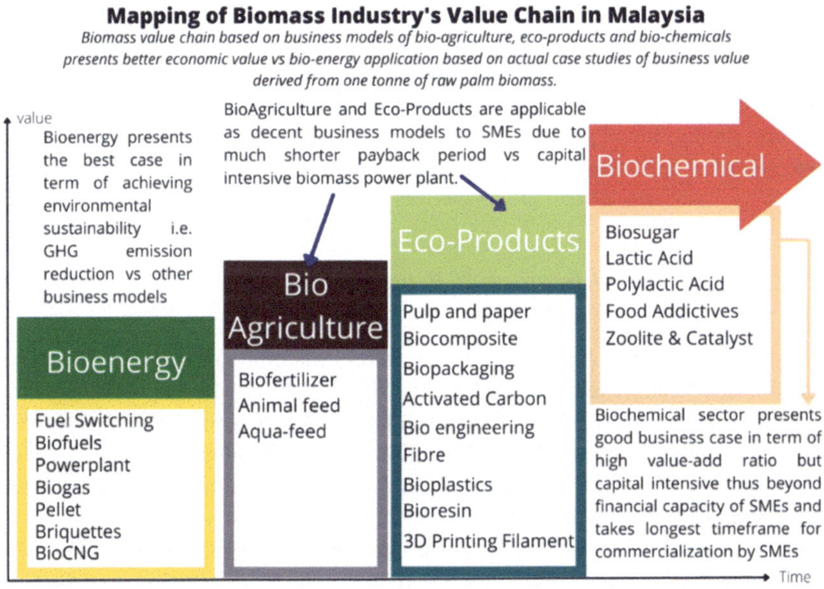

Fig. 3 Mapping of biomass industry's value chain in Malaysia

4 Agro-waste Contents

Fruit and vegetable crop waste such as apple pomace, rinds, skin, shells, peels and vegetable trimmings, inedible dough, solid grains, and discarded portions are categorized as agro-solid wastes. Solid waste contains high levels of carbohydrates, lignin, cellulose, dietary fibres, lipids, and monosaccharides (especially glucose and fructose) (Carol 2020). Cellulose, lignin, and lignocellulose are biodegradable and biocompatible which can be derived from these abundant lignocellulose biomass resources. Lignocellulose mass contains more than 50% hemicellulose and cellulose, which can also be broken down into fermentable pentose (xylose and arabinose) and hexose (glucose) by specific enzymes, respectively. These fermentable sugars are an important substrate for the production of biochemical higher-value products. However, because of the structural complexity, lignocellulose's enzymatic digestibility is weak (Qin et al. 2015). The pretreatment process plays the primary role in the dissociation of biomass structure and increasing the penetrability of enzymes and water to the components. Various pretreatment techniques have been used to enhance lignocellulose's enzymatic digestibility, such as ionic liquid, alkaline, dilute acid, and steam explosion pretreatment (Qin et al. 2017).

This paper focuses on the valorization of coconut fibres (CF) for innovative bioeconomy applications. Malaysians ate the most coconut in terms of weight in 2019 and 2020 with 22.7 and 24.3 kg per person, for each year [18]. In total, Malaysians consumed a total of 790,708,000 kg, or 790 million kg, or 790,708 tonnes of coconuts (Ida, 2021). Coconut fibre consists of 33.20% cellulose, 31.10% hemicellulose, and 20.5% lignin makeup CF (Aziz et al. 2018). CF has a potential cellulose yield of 11.50 million tonnes, according to calculations (Ida, 2021). Nutritional recycling and long-term production of platform chemicals like LA are made possible by the proper management of agro-waste (Thygesen et al. 2021).

4.1 Chemical Modification of Biomass

To boost the enzymatic digestibility of fibres, alkaline pretreatment has been frequently used. However, this method has significant drawbacks, including large chemical loading needs and irreversible chemical adsorption on the surface of processed biomass, reducing the effectiveness of the pretreatment process. To avoid using a large amount of alkali or acidic chemicals, which would complicate the downstream process and raise production costs, the pretreatment process will be carried out with an organic amine (EDA), which has previously been used in the pretreatment of rice straw used in the production of LA (Chen et al. 2019). To eliminate the EDA remains in pretreated biomass and enhance biomass enzymatic digestibility, critical factors and optimal conditions in pretreatment and enzymatic hydrolysis should be studied. In this research, CF was pretreated with EDA under ambient pressure for a constant period. Also, to maximize fermentable sugar conversion, the EDA loading and pretreatment temperature were tested.

The following scope of work has been covered in this paper to fulfil the goal:

(1) Pretreating CF according to the following guidelines: Temperature (90 °C) and EDA concentration (1–10%) for 1 h.
(2) XRD analysis of cellulose I, cellulose III, and amorphous cellulose concentration in pretreated substrates.
(3) FTIR analysis of the pretreated substrates for lignin, cellulose, and hemicellulose content.

5 Materials and Methods

This section deliberates the procedures and materials utilized in the study. A pretreatment step on CF is necessary to effectively utilize these lignocellulosic substrates for further bioconversion.

5.1 Preparation of Coconut Fibre (CF)

The CF was obtained from a local greengrocer mart in Kuala Lumpur, Malaysia. Each substrate was dried in an oven at 60 °C for three days before being milled (0.25 mm) and kept separately in containers for future usage.

5.1.1 Pretreatment of Substrates by EDA Aqueous Solution

The pretreatment process was based on a modified method from the study of Chen et al. (2019). In glass beakers, the substrates were first mixed separately with dilute EDA (1, and 10% v/v) solutions. The solid-to-liquid ratio remained constant at 1:10. Aluminium foil was used to cover the beakers. The mixes were then cooked in an oven at a fixed temperature of 90 °C. The solid was recovered using filter paper without any washing procedures after 1 h of preparation. The substrates were then dried overnight at 60 °C. Finally, the substrates were placed in screw cap test tubes before being submitted for analytical analysis.

5.2 *Analytical Analysis*

5.2.1 Fourier-Transform Infra-Red Spectrum

The cellulose, hemicellulose, and lignin content of untreated and pretreated substrates were determined by Nicolet iS50 FTIR Spectrometer equipped with an Attenuated Total Reflectance (ATR) sampling accessory based on the [25] method. Samples (20 mg) were pressed uniformly against the crystal surface via a spring anvil. Each sample's spectrum was recorded separately by averaging 32 scans collected from wavenumbers of 4,000–400 cm^{-1} at a resolution of 4 cm^{-1}. OMNIC 9.2.86 software was used to record and analyze the FTIR ATR data. The software Origin 7.5 was used to plot the data.

5.2.2 Crystalline Index (CI) and Cellulose Allomorph (CA)

The crystalline index (CI) and cellulose allomorph (CA) had been estimated using a D2 Phaser X-ray Diffractometer. Samples were milled and scanned at a speed of 0.250 time/step, ranging from $2\theta = 10$–40º. The samples were put on a quartz sample holder, and CuKα radiation operated at 30 kV, 10 mA. PeakFit (SeaSolve Software Inc.) software was used to deconvolve the resulting diffractogram. Curve fitting analysis was carried out using Gaussian functions. The content of amorphous cellulose, cellulose I, and cellulose III were calculated by using Eqs. (1–3).

$$\mathrm{I\%} = \frac{\mathrm{A_I}}{\mathrm{A_{AM}} + \mathrm{A_{III}} + \mathrm{A_I}} \times 100\% \tag{1}$$

$$\mathrm{III\%} = \frac{\mathrm{A_{III}}}{\mathrm{A_{AM}} + \mathrm{A_{III}} + \mathrm{A_I}} \times 100\% \tag{2}$$

$$\mathrm{AM\%} = \frac{\mathrm{A_{AM}}}{\mathrm{A_{AM}} + \mathrm{A_{III}} + \mathrm{A_I}} \times 100\% \tag{3}$$

$$\mathrm{CI} = \frac{\mathrm{I_{002}} - \mathrm{I_{AM}}}{\mathrm{I_{002}}} \times 100\% \tag{4}$$

where AM %, III %, and I % are the contents of amorphous cellulose, cellulose III, and cellulose I in samples, respectively; AAM, AIII, and AI are the peak areas of amorphous cellulose, cellulose III, and cellulose I, respectively. The Bragg angles of peak (004), (020), $(1\bar{1}0)$, and (110) belonging to cellulose I: 34.5°, 22.3°, 16.3°, and 14.8°, respectively. The Bragg angles of peak (020) and (110) belonging to cellulose III are 20.0° and 11.3°, respectively. The Bragg angle of the amorphous peak is around 20.5°. The crystallinity index CI was calculated by Eq. (4) (Qin et al. 2018).

6 Results and Discussions

6.1 FTIR Result

The chemical structure was analyzed using FTIR spectra between 4000–400 cm^{-1}. Figure 4 compares the FTIR spectra of raw and EDA-pretreated CF. All the substrates showed a similar peak profile before and after pretreatment. The characteristic absorption peak at 1722 cm^{-1} (corresponding to the C=O of acetyl groups of hemicellulose and ester connecting with lignin) decreased after pretreatment, indicating that a part of lignin has been removed (Qin et al. 2018 and Chen et al. 2019. The absorption peaks at 1399 and 1341 cm^{-1} (corresponding to the stretching vibration of C–N and flexural vibration of C–CH) increased obviously after pretreatment due to the destruction of glucosidic bonds and the bonding of EDA molecules with cellulose and hemicellulose. The increase in band intensity at 1319, 1370 cm^{-1} (representing O–H stretching vibration), and 1424 cm^{-1} in Fig. 4 refers to increased cellulose content due to the removal of lignin and hemicelluloses (Qin et al. 2018).

Among other pretreatment techniques, alkaline pretreatment is one of the most powerful methods that can improve the enzymatic hydrolysis of fibres by disrupting lignin and breaking the bond of lignin carbohydrates (Qin et al. 2017 and Chen et al. 2019). This method also showed the advantages of effectively reducing the crystallization rate by swelling, excluding the acetyl group and uronic acid replacements from hemicelluloses, and breaking down the ester bonds. The main hurdle of inorganic alkaline pretreatment is the release of large amounts of salt from black liquor. To overcome this problem, biomass was pretreated by the organic amine ethylenediamine (EDA) (Chen et al. 2019). Additionally, EDA has been generally used in converting cellulose allomorph as well as ammonia. Cellulose Iβ and cellulose Iα are the official allomorphic forms of cellulose found in plants and microbes, respectively (Qin et al. 2018).

EDA can enter the natural crystalline cellulose as shown in Fig. 4, crack the hydrogen bonds between adjacent cellulose chains, make new hydrogen bonds

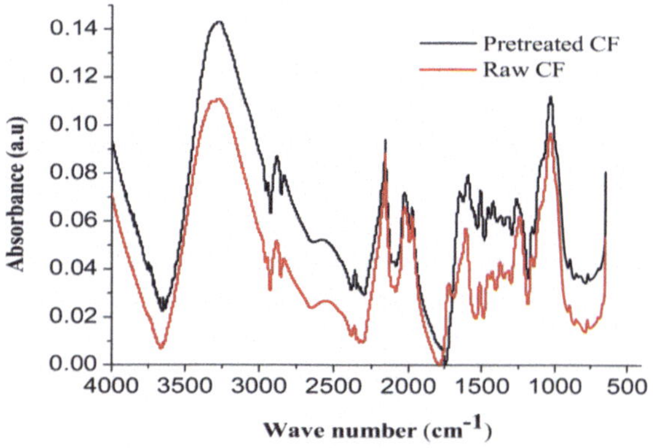

Fig. 4 Comparison of the surface feature of raw and pretreated CF by ATR-FTIR

between cellulose and EDA, and transform cellulose I to EDA–cellulose I complex. After eliminating EDA by drying under vacuum or washing with polar/non-aqueous solutions (e.g., ethanol), the hydrogen bonds rebuild between cellulose units. The crystal configuration of cellulose turns into cellulose IIII. Cellulose IIII can also return to cellulose I by heating above 200 °C (Qin et al. 2018). In a study, after EDA pretreatment, the enzymatic transformation of cellulose was higher than 80%, which may refer to the fact that the enzymatic saccharification ratio of cellulose IIII is much higher than cellulose I. After EDA pretreatment, the majority of the hemicellulose and cellulose stay in a water-insoluble solid form. Lignin is translocated with ether and ester bonds in lignin–hemicellulose complex fraction to the surface of biomass, and incomplete crystal cellulose is converted to amorphous cellulose (Qin et al. 2018).

EDA pretreatment has numerous benefits compared to other methods like dilute acid: it can be performed at ambient pressure, which decreases process cost; it can be carried out without adding water (a dry-to-dry process), which bypasses solid–liquid separation after pretreatment and overcomes energy consumption throughout pretreatment; it protects hemicellulose and does not release fermentation inhibitors like furans. On the other hand, during the pretreatment process, EDA evaporates so it can be reused. Moreover, to lessen refinery costs, EDA should be eliminated from biomass. An earlier study also proposed that EDA loading and pretreatment temperature influenced EDA removal and enzymatic hydrolysis of biomass (Qin et al. 2015).

6.2 XRD Results

The deconvolution of an X-ray diffractogram was used to determine the change of allomorphic forms between different EDA pretreatment settings. The calculated curves were remarkably similar to the observed data (R2 was greater than 0.95 for each sample). Untreated substrate cellulose was found to be cellulose I, with typical peaks (004), (020), (10), and (110) with Bragg angles of roughly 34.5°, 22.3°, 16.3°, and 14.8°, respectively (Fig. 5). With a maximum intensity of around 21°, the amorphous cellulose peak was substantially broader. Untreated CF contained 28.46 cellulose I, and 71.54 p5, amorphous cellulose, as determined by peak areas with crystallinity index, (CI), of 34.52 p5, 24.37 p5, and 20.09 p5, respectively (calculated by Eq. 4). Cellulose allomorph after EDA pretreatment was a mixture of cellulose I and cellulose IIII; cellulose IIII had characteristic peaks (020) and (110) at 20.3° and 11.3°, respectively. The content of cellulose III was influenced by EDA concentration and pretreatment temperature (Table 2). The highest cellulose III content in pretreated CF was around 5% was achieved at 90 °C, 1, and 10% EDA, respectively (Fig. 6). The CI increased from 34.52% up to 36.79% in CF after EDA pretreatment. The increase of CI could be due to the removal of amorphous components (hemicellulose and lignin) by EDA pretreatment (Xu et al. 2020 and Zhu et al. 2018). The ratio of CI to cellulose content is regarded as an appropriate means to estimate actual crystallinity in native cellulose (Xu et al. 2020 and Zhu et al. 2018) The CI/cellulose ratio decreased from 1.21 in raw CF to 0.62. It illustrated that while the crystallinity of the pretreated CF increased due to the removal of amorphous components, the crystallinity of cellulose itself decreased as the cellulose content dramatically increased.

Based on these findings, the pretreatment of CF under the following parameters: EDA concentration (1–10%) at the temperature 90 °C for 1 h compared to untreated CF had shown significant changes in the content of cellulose I, cellulose III, and amorphous cellulose in the pre-treated substrates by XRD. The identification of lignin, cellulose, and hemicellulose content in the pretreated and untreated substrates by FTIR showed significant modifications had been made by EDA.

Table 2 Crystalline cellulose, amorphous cellulose, and crystallinity index contents of untreated and pretreated CF at different pretreatment conditions

EDA (%)	Temperature (°C)	Coconut Fiber (CF)				
		I (%)	III (%)	AM (%)	CI (%)	CI/cellulose
10	90	45.97	1.29	52.74	39.37	0.83
1	90	54.05	4.82	41.13	36.79	0.62
Untreated		28.46	0.00	71.54	34.52	1.21

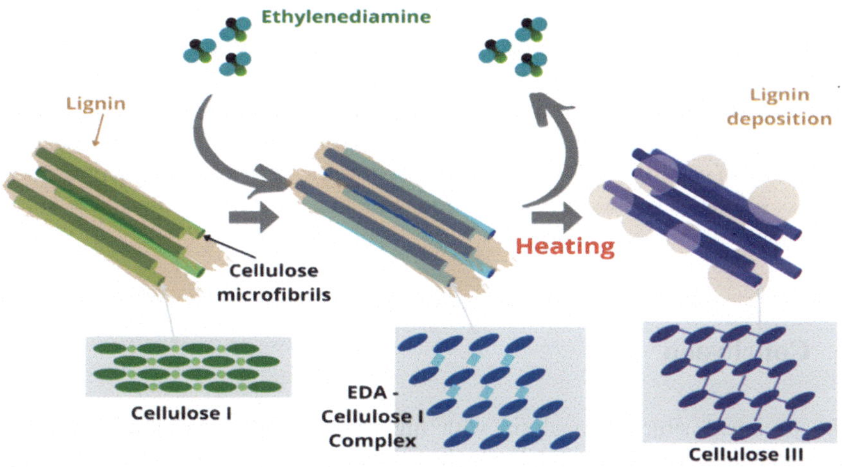

Fig. 5 Approximate schematic for lignocellulosic biomass pretreated by EDA

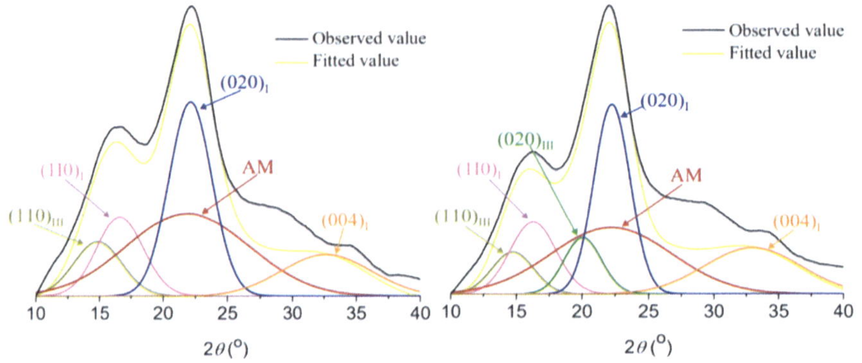

Fig. 6 X-ray diffraction profiles. **a** Untreated CF and pre-treated CF at 90 °C, 1% and 10% EDA, respectively

6.3 Applications and Prospects

Using lignocellulose, particularly lignocellulose from agricultural or forestry waste, has the advantage of minimizing waste in the environment while also enhancing the value of the waste. Agro-wastes pretreatment process using green solvents had assisted the transformation of complex starch molecules in cellulose to be converted into fermentable sugars which will be beneficial for the microorganisms to harness them into LA. Bio-based LA production, particularly using cheap carbon sources with agro-waste is getting more attention. The main bottlenecks are the low production yield and the inconsistency of the biopolymers. Chemical modifications together with cost-effective downstream processing are promising approaches to overcome the hurdles of commercial PLA production from waste streams. Thus, a pretreatment process similar to this work is required for fungi and bacteria to effectively utilize these lignocellulosic substrates.

7 Conclusion

The chemical structure analysis with high lignin and hemicellulose removal supported the preliminary study's conclusion that EDA pretreatment had drastically degraded the physically structural wall of the CF. This preliminary finding is beneficial in assisting the bioeconomy through the agro-waste valorization process as a tremendous amount of agro-waste is produced daily. To resolve these problems, implementation of a chemical modification of starch, cellulose, and lignin at the preparation platform for agro-waste is essential. The adoption of such a process can produce value-added products while reducing waste as well as economically feasible.

Acknowledgements The authors would like to thank the Ministry of Higher Education, Malaysia for the Fundamental Research Grant of FRGS21-251-0860 and Department of Chemical Engineering and Sustainability of Faculty Engineering, International Islamic University of Malaysia for the facilities and equipment.

References

Abdulrazik, A., Elsholkami, M., Elkamel, A., & Simon, L. (2017). Multi-products productions from Malaysian oil palm empty fruit bunch (EFB): Analyzing economic potentials from the optimal biomass supply chain. *Journal of Cleaner Production, 168*. https://doi.org/10.1016/j.jclepro.2017.08.088

Akolgo, G., Awafo, E., Essandoh, E., Owusu, P., Uba, F., & Adu-Poku, K. (2021). Assessment of the potential of charred briquettes of sawdust, rice and coconut husks: Using water boiling and user acceptability tests. *Scientific African, 12*, e00789. https://doi.org/10.1016/j.sciaf.2021.e00789

Aziz, S., Shariff, A., Abdullah, N., & Mohamed Noor, N. (2018). Characteristics of coconut frond as a potential feedstock for biochar via slow pyrolysis. *Malaysian Journal of Fundamental and Applied Sciences, 14*, 408–413. https://doi.org/10.11113/mjfas.v1.

BGB. (2021). *BGB's commitment to the circular economy*. Retrieved from BGB: https://www.bgbinnovation.com/circulareconomy/.

Carol, S. K. L., Guneet, K., Chong, L., Xiaofeng, Y., & Christians, V. S. (2020). *Waste valorisation: Waste streams in a circular economy*. Wiley.

Chen, H., Weizhong, H., Wang, B., Wang, Y., Wen, H., Changwei, Z., Wu, Y., & Qin, P. (2019). L-lactic acid production by simultaneous saccharification and fermentation of dilute ethylediamine pre-treated rice straw. *Industrial Crops and Products., 141*, 111749. https://doi.org/10.1016/j.indcrop.2019.111749.

Malode, S. J., Prabhu, K. K., Mascarenhas, R. J., Shetti, N. P., & Aminabhavi, T. M. (2021). Recent advances and viability in biofuel production. *Energy Conversion and Management, X*(10), 100070. ISSN 2590-1745. https://doi.org/10.1016/j.ecmx.2020.100070.

Qin, L., Li, X., Zhu, J.-Q., Iwinchy, L., Xu, H., Guan, Q.-M., Zhang, M.-T., Li, B.-Z., & Yuan, Y.-J. (2017). Optimization of ethylenediamine pretreatment and enzymatic hydrolysis to produce fermentable sugars from corn stover. *Industrial Crops and Products, 102*, 51–57. https://doi.org/10.1016/j.indcrop.2017.03.026.

Qin, L., Zhao, X., Li, W. C., Zhu, J. Q., Liu, L., Li, B. Z., & Yuan, Y. J. (2018). Process analysis and optimization of simultaneous saccharification and co-fermentation of ethylenediamine-pretreated corn stover for ethanol production. *Biotechnology for Biofuels, 11*, 118. https://doi.org/10.1186/s13068-018-1118-8.

Qin, L., Li, W. C., Zhu, J. Q., Liang, J. N., Li, B. Z., & Yuan, Y. J. (2015). Ethylenediamine pretreatment changes cellulose allomorph and lignin structure of lignocellulose at ambient pressure. *Biotechnology for Biofuels, 8*, 174. https://doi.org/10.1186/s13068-015-0359-z.

Reike, D., Vermeulen, W., & Witjes, S. (2017). The circular economy: New or Refurbished as CE 3.0?—Exploring controversies in the conceptualization of the circular economy through a focus on history and resource value retention options. *Resources, Conservation and Recycling*.https://doi.org/10.1016/j.resconrec.2017.08.027.

Teysha. (2018). TeyshaTech. (SIRIM) Retrieved 29 December, 2023, from https://teyshatech.co.uk/sirim-bioplastics-pilot-plant-malaysia/.

The Center for International Forestry Research (CIFOR) and World Agroforestry (ICRAF). (2021). Retrieved 30 November, 2023, from https://www.cifor.org/wp-content/uploads/2021/03/Flyer%20-%20Knowledge%20Guide_Circular%20Bioeconomy-v4.pdf.

Thygesen, A., Tsapekos, P., Alvarado-Morales, M., & Angelidaki, I. (2021). Valorization of municipal organic waste into purified lactic acid. *Bioresource Technology, 342*, Article 125933. https://doi.org/10.1016/j.biortech.2021.125933.

Xu, L., Iwinchy, L., Xin, J., Zhang, S.-J., Li, B.-Z., & Yuan, Y.-J. (2020). Ethylenediamine enhances ionic liquid pretreatment performance at high solid loading. *ACS Sustainable Chemistry & Engineering, 8*, 13007–13018. https://doi.org/10.1021/acssuschemeng.

Zareei, S. A., Ameri, F., Dorostkar, F., & Ahmadi, M. (2017). Rice husk ash as a partial replacement of cement in high strength concrete containing micro silica: Evaluating durability and mechanical properties. *Case Studies in Construction Materials, 7*. https://doi.org/10.1016/j.cscm.2017.05.001.

Zhu, J. Q., Wu, X. L., Li, W. C., Qin, L., Chen, S., Xu, T., Liu, H., Zhou, X., Li, X., Zhong, C., Li, B. Z., & Yuan, Y. J. (2018). Ethylenediamine pretreatment of corn stover facilitates high gravity fermentation with low enzyme loading. *Bioresource Technology, 267*(2), 227–234. https://doi.org/10.1016/j.biortech.2018.07.030. Epub 2018 Jul 7.

Biomaterials

Agricultural Waste Fibers Reinforced Polystyrene Composite

Yusilawati Ahmad Nor, Mahirah Sarah Mansor, Syazana Ab Manaf, and Nur Ayuni Jamal

1 Introduction

Polystyrene (PS) is a non-degradable plastic made up of styrene polymer which is a synthetic resin. PS is widely used in many fields such as packaging and building insulation. A huge global production of this material is around 15.61 million metric tons in 2019 and it is expected to increase gradually to 15.68 million metric tons per year by 2024 (Garside 2020). The high amount of PS production will cause abundant accumulation of plastic waste and the incineration of this material causes the release of harmful chemicals that can cause cancer in human. They can cause major land and ocean pollution due to its non-degradable and non-recyclable properties (Aciu et al., 2015). Most of the PS that ends up in landfills will still remain there approximately 500 years (Little Mark 2021) with approximately 8 million tons of these plastic wastes escape into the oceans every single year. The pollution is mostly visible in developing countries, where garbage collection systems are often inefficient with low recycling rates. Table 1 below shows the plastic categories and the recyclable status of plastic in Malaysia. It is shown that PS is not recyclable as compared to polyethylene terephthalate, high-density polyethylene and polypropylene which can be recyled.

Several efforts have been made to improve degradability and reusability of the PS by reinforcing it with agricultural waste. Agricultural waste is unwanted or unsalable materials produced wholly from agricultural operations directly related to the growing of crops or raising of animals for the primary purpose of making a profit or

Y. A. Nor (✉) · M. S. Mansor · S. A. Manaf
Department of Chemical Engineering and Sustainability, Faculty of Engineering, International Islamic University Malaysia, Selangor, Malaysia
e-mail: yusilawati_ahmadnor@iium.edu.my

N. A. Jamal
Department of Manufacturing and Material Engineering, Faculty of Engineering, International Islamic University Malaysia, Selangor, Malaysia

F. Adyani Ahmad Fuad et al. (eds.), *Circular Bioeconomy*,
https://doi.org/10.1007/978-981-97-7010-6_10

Table 1 Category of plastics and recyclable status in Malaysia (Khazanah Research Institute 2019)

No	Category	Product of uses	Recyclable in Malaysia
1	Polyethylene terephthalate (PET/PETE)	Mineral water bottles, cookie jars	Yes
2	High-density polyethylene (HDPE)	Milk containers, buckets, shampoo bottles	Yes
3	Polyvinyl chloride (PVC)	Pipes, synthetic leather	No
4	Low-density polyethylene (LDPE)	Bubble wrap, plastic bags	No
5	Polypropylene (PP)	Disposable food containers, bottle caps	Yes
6	Polystyrene (PS)	Disposable cups, plates, cutlery	No
7	Others	Miscellaneous plastic, nylon	No

for a livelihood. Agricultural waste has been selected as composite material due to its high strength fiber, environmentally friendly and high availability thus contributes to the low cost of composite process. Furthermore, about 1.2 million tons of agricultural waste is disposed annually into landfill and it is estimated that the total agricultural waste generation in Malaysia can reach up to 0.210 kg/capital/day by 2025 (Tahir 2012). Few of the examples include sugarcane bagasse, fruit trees, vegetables, and date palm. It is predicted that the composite material will be a good material in various applications for instance, in non-structural construction, secondary food packaging industry, and art industry. The effort on the development of the composite by compositing both materials will reduce the waste generation. Therefore, it is the focus of this chapter to review on the most recent studies of the topic.

2 Polystyrene Properties and Applications

Polystyrene (PS) can be hard, solid plastic, or in foam state. There are three common types of PS which include PS foam, regular PS plastic, and PS film. For PS foam, it has been divided into two which are expanded polystyrene (EPS) and extruded polystyrene (XPS). EPS is the most common type that usually used for food packaging. XPS is a high-density foam commonly used in application like architectural building model. PS plastic has been divided into two which are homopolymer and copolymer. Homopolymer PS is brittle and can be made more impact resistant by combining with other materials such as rubber which is known as copolymer high impact polystyrene (HIPS). Moreover, polystyrene film is also one of the PS types that can be in vacuum form and used in packaging applications. Films can be made into oriented polystyrene (OPS) that is cheap to produce, however it is a brittle material.

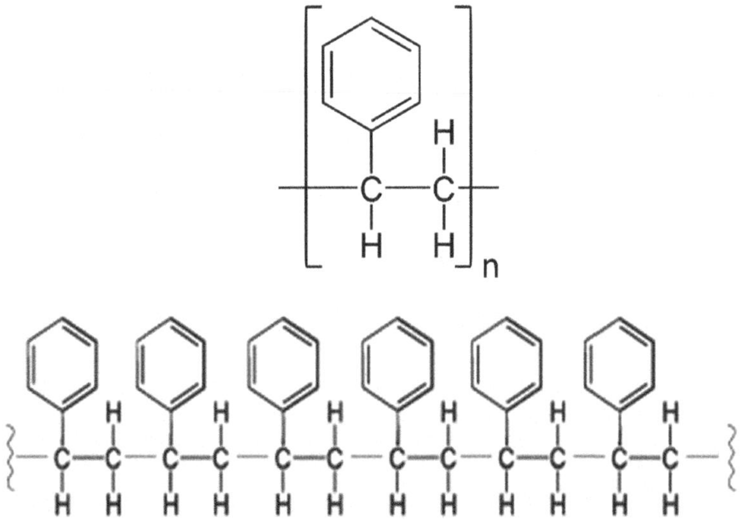

Fig. 1 The chemical structure of PS (PubChem 2022)

According to Collier County (2020), PS is non-degradable and takes approximately 500 years to degrade in environment. PS is an aromatic hydrocarbon, and it has hydrophobic surface which is resistant to water due to the presence of methane ($-CH_3$) or ethylene ($-CH=CH_2$) on its surface. Every alternate carbon atom on the PS chain has a phenyl group which is a benzene ring attached to it. It is a long-chain hydrocarbon with chemical formula is $(C_8H_8)n$. The chemical structure of PS is shown in Fig. 1 below.

PS has a low glass transition temperature (tg) which is approximately at 100 °C, (SpecialChem 2020). It is flammable due to the properties of ethylene and benzene itself are flammable. The material will start releasing flammable gas when the operating temperature reaches approximately 200 °C. PS is low in thermal conductivity, thus providing excellent thermal insulation. The thermal conductivity of polystyrene is reported around 0.030 to 0.040W/m•K, (MatWeb 2020a 2020b).

Although it is low in impact strength which is around 15–20 MPa due to the stiffness of the polymer backbone, it has medium to high tensile strength and tensile modulus with value 35 MPa up to 55 MPa and 2.91 GPa, respectively, (MatWeb 2020a 2020b). Moreover, expanded PS is high in compressive strength and very useful for heavy load bearing application. It is reported that the tensile modulus, tensile strength, and compression strength are 51 MPa, 1.1 MPa, and 40 MPa respectively at 1100 kg/m^3 density, (MatWeb 2020a 2020b). PS is a waterproof material and resistant to aqueous solution of acid and alkaline, however it is not resistant to organic solvents. The properties of polystyrene were summarized in Table 2.

Table 2 Properties of polystyrene

Properties	Description
Non-degradable	About 500 years to degrade in environment
Hydrophobic	Presence of methane(–CH_3) or ethylene (–$CH{=}CH_2$) at its surfaces
Strong and durable	High tensile strength and tensile modulus which around 35 MPa to 55 MPa and 2.91 GPa, respectively
Ignition temperature	200° C
Glass transition temperature (tg)	100 °C
Low in thermal conductivity	The diffusivity is around 0.03–0.04 K

3 Agricultural Waste and Its Properties

Many types of agricultural waste are available to be used as the reinforcement material, which include rice husk, wheat husk, coconut shell, and curuau fiber. Rice husk, for instance, is the hard protective covering of rice grain. It is usually separated from the rice grains during the milling process to yield white rice. The common rice growing regions are China, India, Japan including in Malaysia. The current production of white rice is estimated around 700 million tons around the world (Alyamac et al. 2018). Similarly, many other agricultural sectors contribute to Malaysia's development such as palm oil, paddies, sugarcane, and other planted crops. According to Wan Basri and his colleagues, Sugarcane bagasse fiber is one of the most rising cash crop in Malaysia with total productions of 28.1 million metric tons in 2016 and 28.0 million metric tons in 2017 (Wan Basri et al. 2019). Gigantic amount of agricultural waste after reaping off its filling or after crushing process is usually left to decay on the fields. The accumulated agricultural waste without proper management will potentially lead to the spreading of diseases that adversely affect humans, animals, and the environment.

Agricultural waste is a type of lignocellulosic biomass that usually consists of cellulose, hemicellulose, and lignin with percentage 50%, 25%, and 25% respectively. The hemicelluloses are made up of C5 and C6 sugar. These biological polymers, hemicellulose and cellulose have phenolic group which are carboxylic(–COOH) and hydroxyl(–OH) groups that contribute to hydrophilic of the surface. In contrast, lignin is not a sugar. It is various in mass and methoxyl (–OCH_3) content where is more to hydrophobic properties. The crosslinking of lignin polysaccharides is an obstacle for water absorption to the cell wall. Cellulose is the major composition in agricultural waste as it would contribute to the major effect of the composite mechanism. It is a tough fibrous and water insoluble polysaccharide, long microfibrils bonded to each other but hydrophilic due to the presence of hydroxyl (–OH) groups on its surface. Microfibrils are bundles of polysaccharide that arranged in fibrils contain about 20 cellulose chains that make the biomaterial with high strength. With higher cellulose content, the biomass fibers derive its strength from hydrogen bonding in the microfibrils of cellulose, thus suitable to be as the reinforcement of a composite material, (Rahman et al. 2017). Figure 2 shows the structure of cellulose that contains of the monomer of glucose joined by β-1,4 linkages.

Table 3 below shows the composition of agricultural waste that mostly consists of cellulose, followed by hemicellulose and lignin. For instance, wheat straw fiber has up to 40% of cellulose, 25% of hemicellulose, and 20% of lignin.

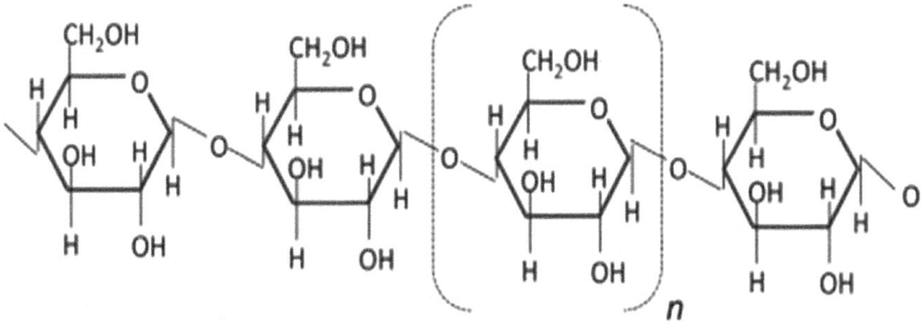

Fig. 2 Structure of cellulose with β-1,4 linkages (Brigham 2018)

Table 3 Composition of agricultural waste (Jahirul et al. 2012)

Agricultural waste	Lignin (%)	Cellulose (%)	Hemicellulose (%)
Wood	25–30	35–50	20–30
Wheat straw	15–20	33–40	20–25
Switch grass	5–20	30–50	10–40
Sugarcane bagasse	23–32	19–24	32–48
Miscanthus	17	24	44
Corn stover	16–21	28	35
Hazelnut shell	42.9	28.8	30.4
Olive husk	48.4	24	23.6
Corncob	15	50.5	31
Tea waste	40	30.20	19.9
Walnut shell	52.3	25.6	22.7
Almond shell	20.4	50.7	28.9
Sunflower shell	17	48.4	34.6
Nut shell	30–40	25–30	25–30
Paper	0–15	85–99	0
Rice straw	18	32.1	24
Stored refuse	20	60	20
Leaves	0	15–20	80–85
Cotton seed hairs	0	80–95	5–20

Table 4 Comparison between sugarcane bagasse and polystyrene properties

Description	Polystyrene	Agricultural fiber
Degradability	Non-degradable	Biodegradable
Tensile strength, TS	35.0 MPa up to 55.0 MPa	20.0 MPa up to 200.0 MPa
Tensile modulus, MOE	2.9 GPa	19.7 GPa
Ignition temperature	200 °C	520 °C
Chemical resistance	Not resistant to organic solvent	Resistant to organic solvent
Transitional temperature, tg	75 to 80 °C	220 °C

Agricultural waste fiber is a biodegradable material, suitable as a reinforcement for PS to improve composite properties. Furthermore, the high availability of agricultural waste fiber will lead to a low cost to obtain the material. It is also high in tensile strength and tensile modulus (Kumar et al. 2014). Its tensile strength from 20 MPa could reach to 200 MPa with a density of 1.2 g/cm^3 while its tensile modulus is 19.7GPa which is quite higher than mechanical properties of PS (Jayamani et al. 2020). Its ignition temperature and transitional temperature (tg) are also higher than polystyrene which are 520 °C and 220 °C respectively (Khatib & Negim 2013). Furthermore, it is also resistant to organic solvent (Falca et al. 2019). By considering all these advantages, agricultural waste has been among the preferred waste material chosen to be used as composite material. Table 4 shows the comparison between PS and Sugarcane bagasse waste properties.

The composition of agricultural waste with polystyrene is a great interest for developing new material due to its biodegradability and reinforcing properties. Agricultural waste fibers have low density, high mechanical resistance, and abundance in nature. Moreover, recycling of polystyrene usually results in partial changes of the material due to the environmental conditions like humidity, temperature, and UV sunlight that eventually contribute to color changes and lost in mechanical properties, (Zizumbo et al. 2011). Thus, by incorporating it with the natural fibers will possibly improve their mechanical and thermal properties.

4 Agricultural Waste Reinforced Polystyrene Composite

Many types of agricultural wastes have been studied in recent years to enhance the properties of PS composite which includes; date palm leaflet, curaua fiber, rice husk, wheat straw, mustard husk, sugar palm fiber, elephant grass, coconut shell, corn husk, and durian husk. Table 5 summarized previous studies of the agricultural waste-polystyrene composite along with the manipulated variables, method of preparation, and the observation in terms of mechanical and its thermal properties. The types of agricultural waste which were commonly used most recently were rice husk (RH) followed by wheat husk (WH) fibers, sawdust, corn husk, and coconut shell. Compared to wheat straw, rice husk fiber has lower reactivity in terms of thermal properties. Meanwhile, the higher reactivity of wheat straw is attributed to higher ash contents and instability of its morphology and structure (Pottmaier et al. 2013).

4.1 General Method for the Preparation of Agricultural Waste Reinforced Polystyrene Composite

Common methods used to prepare the agricultural waste reinforced polystyrene composite include solution casting or the melt blending method using either compression molding, injection molding, thermoforming, or extrusion techniques. Prior to the composite preparation, agricultural fiber would be sieved depending on the required size and further pre-treated either with an alkaline or acidic solution. For alkaline treatment, NaOH can be added to the fiber at 120 °C by mixing in a beaker followed by washing it using distilled water and dry heating. The fiber would also be directly washed with distilled water only followed by dry heating of the fiber without the pre-treatment process. For solution casting method, PS would be immersed in solvents such as acetone, toluene, benzene, gasoline, or chloroform solution at room temperature to obtain smooth paste. Later, fibers and melted PS are homogeneously mixed or blended at a temperature above 110 °C. Coupling agent, processing aid, or fire retardant could be added together if it is considered in the process. This mixture would be poured directly into a mold or pressed to produce the samples at the desired dimension. The samples are later cured in air or oven for few hours before being used for final application such as green packaging or insulator depending on the final aims. For the melt blending method, PS would be first melted in an oven before being compounded with the fiber in a mixer at 190 °C. The composite would be then molded into its final form using either the injection molder, extruder (i.e., twin screw extruder), or compressor (Ling et al. 2019). Other less common method includes in-situ polymerisation of the fiber and monomer (styrene) mixture to form the composite with the presence of an initiator (Ghorbani et al. 2018; Zafar & Siddiqui 2020a 2020b).

4.2 Parameters Affecting the Properties of Agricultural Waste-Polystyrene Composite

The mechanical properties of the composite material such as tensile strength (TS), modulus of rupture (MOR), and modulus of elasticity (MOE) were improved depending on several variables that were manipulated during the processing of the composite material and are highly dependent on the characteristic of the agricultural waste filler as the reinforcement material. These factors include; (1) diameter of fiber, (2) length of fiber, (3) fiber loading concentration (4) morphology of fiber, and (5) pre-treatment of fiber. In addition, the usage of coupling agent, processing aids, and nanoparticles such as multi wall carbon nanotubes (MWCNTs) will also

Table 5 Summary of the preparation of agricultural waste reinforced polystyrene composite

Types of agricultural wastes	Manipulated variables	Method of preparation	Observation	References
Date palm leaflet (LPC)	Size of fiber	PS was dissolved in gasoline. LPC was sieved based on different mesh size. LPC and PS were mixed together and poured into compression mold to form a sample	As the particle size of fiber increases, tensile strength (TS) and MOR value of composite material decreases	Masri et al. (2018)
Curaua fiber (CF)	Morphology of fiber Nanofiber (CNF) vs Nano crystal (CNC)	PS was solubilized by using acetone. CNFs were obtained by mechanical defibrillation while CNCs were obtained by (T.E.M.P.O.)-mediated oxidation method. Ethanol was added to CNFs or CNCs for solvent exchange. CNFs or CNCs were mixed together with PS and compatibilizer by using twin screw extruder. The compound was molded by using injection machine	Incorporation of CNF promoted an increased in the compressive strength of composite materials compared to CNC and PS alone. However Tg value of the composite materials were lower than PS	Neves et al. (2019)
Rice straw (RC)	Addition of coupling agent	Rice straw was pre-treated by using sodium bicarbonate. Metal solution were prepared by adding iron (Fe^{3+}) and nickel (Ni^{2+}). The metal solution was mixed with coupling agent, cobalt(CO^{2+})(Co(ii)). Lignin was isolated from fiber by mixing fiber together with metal solution. The samples were extruded with PS by using twin screw at 130 °C. The samples were molded into films using compressor machine	TS of composite material with coupling agent was higher than PS alone. However, tg of the composite materials were lower than PS	El-Zawawy et al. (2011)

(continued)

Table 5 (continued)

Types of agricultural wastes	Manipulated variables	Method of preparation	Observation	References
Wheat straw (WS)	Types of pre-treatment:	Wheat straw was sieved and treated by using either hydrochloric acid or sodium hydroxide. The composite was prepared using solution casting methods. PS was dispersed in toluene. Treated wheat straw and polybutene plasticizer were added subsequently to PS solution. The compound was cast in a petri dish	TS and Tg value of alkaline-treated composite were higher than acid-treated composite due to better adhesion with the PS matrix	Dixit and Yadav (2020)
	Alkaline vs acidic		However, Tg of all composite materials were lower than PS	
Pineapple leaf (PL) fiber	Modification of the cellulose fiber:	The fiber from pineapple leaf was modified using selected method (OF, SF, and GF) before being blended with the PS in a counter rotating internal mixer at 160 °C to form granules. The compounds were later shaped in compression molding	Composite materials have higher TS than PS	Shih et al. (2020)
	(1) TEMPO radical oxidized (OF)		Composite reinforced with SF modified PL fiber has the highest TS followed by GF and OF modified fiber	
	(2) Suspension polymer-ization (SF)		However, Tg of all composite materials were lower than PS	
	(3) Sol–gel modified (GF)			

(continued)

Table 5 (continued)

Types of agricultural wastes	Manipulated variables	Method of preparation	Observation	References
Sawdust (SW) and Rice Husk (RH) fiber	Fiber concentration, fiber size, and alkaline treatment concentration	Fibers were dried, sieved, and treated with NaOH. PS was dissolved in petroleum solvent. Fibers and PS were mixed in a container. The compound was poured into a mold to obtain the sample	High impact strength of the composite was achieved when fiber concentration was at 15% for both fibers, fiber size of 100 um and 31 um for SW and RH respectively and 4% and 2% alkali concentration for SW and RH respectively. Lower fiber size and higher fiber concentration give high impact on mechanical strength of the composite	Abdulkare et al. (2019)
Sawdust (SD) fiber	Size of fiber	Sawdust was sieved at sizes 0.85 and 1.75 mm. PS was dissolved in gasoline. PS and sawdust were manually mixed together. The mixture was poured into two wooden molds. The mold were cold pressed in a hydraulic jack press. The samples were oven-dried at 150 °C	Modulus of rupture (MOR) and modulus of elasticity (MOE) of smaller size sawdust fiber reinforced composite is higher than PS alone, while composite with larger size of sawdust fiber was lower than PS	Akinyemi et al. (2019)
Rice husk (RH) fiber	Colloidal silica extracted from RH and composite preparation (nanosilica alone, nanosilica + PS, PS)	Composite was prepared using vacuum/pressure method with nano silica suspension and styrene monomer. Styrene polymerisation was done through thermal catalyst treatment	MOE and MOR of composite materials were higher than PS alone	Ghorbani et al. (2018)

(continued)

Table 5 (continued)

Types of agricultural wastes	Manipulated variables	Method of preparation	Observation	References
Wheat husk, rice husk, mustard husk	Types of fibers, fiber size, fiber loading	Styrene was added with NaOH to remove inhibitors. It was heated up to 110 °C for polymerisation with addition of benzoyl peroxide as initiator. Fiber particles were mixed in a beaker when polymerisation started. It was transferred from beaker to plastic plate and kept in oven at 120 °C to get solid polymer. The solid polymer was poured into injection molding to prepare different sample sizes	TS and MOR of all composite materials were higher than PS Wheat husk has the highest TS, while Rice Husk has the highest MOR. Fiber size of 250–355 um and 5% fiber loading give the maximum values for tensile strength, flexural strength and hardness of the composite	Zafar and Siddiqui (2020a)
Rice straw (RS) fiber	PS concentration and electron beam irradiation dose	PS was dissolved by using trichloro-ethylene solvent. RS fibers and PS were mixed in a beaker. The mixture was pressed in a steel mold using an electric hot press at 130 °C to form the composite	MOR and MOE increased when PS percentage and electron beam radiation dose increases in the composites	Ismail et al. (2011)
Sugar palm (SP) fiber	Fiber concentration	SP fiber and PS were mixed together by using a melt mixer at 165 °C. The compound was placed in compression mold to obtain specimen	TS value of the composite reduced with increasing fiber loading MOE value of the composite increased with fiber loading concentration	Sapuan et and Bachtiar (2012)

(continued)

Table 5 (continued)

Types of agricultural wastes	Manipulated variables	Method of preparation	Observation	References
Rice straw (RS) fiber	Addition of coupling agent (PS-g-MA) and different fire retardants (zinc borate & magnesium hydroxide)	PS was dissolved in acetone. EPS and coupling agent (PS-g-MA) were preheated. Dried RS was added gradually to the molten PS. Fire retardant (MH) was added to the molten. The compound was pressed by using hydraulic hot press for the testing sample	TS and Tg values are higher when coupling agents and fire retardants were added in the composite compared to PS alone or PS with fire retardant only	Eskander et al. (2018)
Elephant grass (EG) fiber	Fiber concentration	PS was dissolved in gasoline. Dried elephant grasses were fed into carbonization unit to yield char. PS and char were manually mixed. The mixtures were poured into mold for sample	Composite materials were higher in heat capacity compared to PS The higher composition of biochar has higher heat capacity as compared to the lower composition of biochar	Adeniyi et al. (2020)
Palm leaves	Types of plastic Polystyrene (PS) Polycarbonate (PC)	Palm leaves were shreded approximately 4.0 mm. PS was dissolved in chloroform. PS and palm leaves were mixed together at ratio 1:1. The mixture was pressed with hydraulic compression under hot molten	PS-mix exhibited higher TS and MOE than the PC-mix	Binhussain and El-Tonsy (2013)

(continued)

Table 5 (continued)

Types of agricultural wastes	Manipulated variables	Method of preparation	Observation	References
Rice husk (RH) fiber	Fiber concentration	PS was dissolved in various solvent. The sieved rice husk was mixed with PS using mixer. The mixture was pressed using roller on metal plate	MOE of composite materials were higher than PS alone	Victor (2020)
			Higher fiber loading percentage was higher in MOE value as compared to lower filler loading percentage	
Coconut shell powder (CSP)	Powder concentration and alkaline pre-treatment	CSP was treated with NaOH. PS was dissolved in acetone. CSP was mixed together in PS solution using stirrer. The compound was poured into a metal plate to be pressed to form a sample	Composites with alkaline-treated CSP have better TS, MOE, and MOR value compared to the one untreated and increased with increasing CSP concentration	Pao and Chan (2018)
Corn starch	Filler pre-treatment:	Starch was stirred in a reactor with hydrogen peroxide and NaOH for modification. PS, corn starch, zinc octanoate, and glycerol were extruded together at six heating zones. The compound was compressed at 180 °C to obtain the film	TS, MOE, and Tg of unmodified corn starch were higher than the modified corn starch	Gutiérrez and Alvarez (2017)
	Native corn starch vs Oxidized corn starch		Tg of composite materials were lower than PS	

(continued)

Table 5 (continued)

Types of agricultural wastes	Manipulated variables	Method of preparation	Observation	References
Sawdust (SD)	SD powder concentration, PS concentration, egg shell powder concentration	Sawdust was sieved to have 0.03 mm diameter. PS was dissolved in hydrocarbon solution and sawdust and eggshell were mixed together manually. The compound was pressed by using cold press for the sample	Composite with SD has reduced TS value and thermal conductivity compared to PS alone and this value decreased with increasing SD concentration. Addition of eggshell improved the TS properties of the composite	Shuaib-Babata et al. (2022)
Rice husk (RH)	Addition of fire retardant (DAHP) and nanoparticles (MWCNTs)	Rice husk (RH) was heated at 700 °C to form ash. Ash from RH was washed with DI water followed by drying at 150 °C. PS was dissolved in chloroform. PS was grounded with (DAHP) or (DAHP and RH) or (DAHP, RH, and MWNTs) using ultrasound. The sample were molded at 150 °C	Composite with RH ash and addition of fire retardants and MWCNTs has the highest Tg compared to the other type of combination	Attia et al. (2015)
Corn husk fiber (CHF)	Fiber pre-treatment and fiber concentration	PS was dissolved in acetone. CHF was sieved using 100 μm mesh. CHF was either treated with sodium hydroxide solution or was added directly into PS solution. PS and CHD were compounded at 190 °C. It was molded into sheet form for sample	Composite with alkaline-treated fiber has better TS, MOE, and tg value as compared to untreated fiber and PS alone. These values increased with increasing fiber concentration	Chun et al. (2018)

(continued)

Table 5 (continued)

Types of agricultural wastes	Manipulated variables	Method of preparation	Observation	References
Durian husk fiber (DHF)	Fiber concentration:	PS was dissolved in acetone. DHF was sieved using 600 μm mesh and later mixed with Ultra-Plast WP516 aid. DHF and PS were mixed using a melt blending process. The compound was compressed and molded	Composite material with aid at a high concentration of fibers has higher TS, MOE, and tg compared to PS alone, while composite material without aid has lower TS and MOE compared to PS	Koay et al. (2018a 2018b)
	15%		Increase of fiber content will enhance TS and MOE properties of composite material	
	60%			
	With aid or			
	Without aid			
Coconut shell (CS) powder	Coupling agent and processing aids:	PS was melted by using oven and later was mixed with MAPS and Ultra-Plast WP516 aid until completely molten. CS was later added and compounded by using mixer at 190 °C. The composite was molded into a square sheet using compression molding machine	TS, MOE, and tg of composite with coupling agent and processing aid were higher than PS/CS composite	Ling et al. (2019)
	PS, Maleated PS (MAPS), Ultra-Plast WP516			

influence the composite properties. Based on Table 5, composite with the agricultural waste fiber of small diameter (250–335 um) would enhance the composite properties with an increment of TS values up to 40% compared to fiber with much bigger diameter (>355 um). This is due to smaller fiber size as a filler has better dispersion throughout the composite material, thus increase the wettability of the fiber particles to the polystyrene matrix (Zafar & Siddiqui 2019). Moreover, smaller fiber particle size will contribute to a higher total surface area to interact with PS matrix, thus better in enduring any tensile or flexural stress, (El-Zawawy et al. 2011).

Composite with a higher length of agricultural waste fiber would also enhance the composite mechanical properties. Neves et al. (2019) found that the compressive strength of the composite increased when curau nanofiber was employed in the composite as compared when nanocrystal structure of the agricultural waste was used. This observation can also be explained due to the morphology of the agricultural waste fiber itself. This is because long fiber has a higher subcritical length compared to nanocrystal. When full tensile stress is applied, it could be totally supported on composite with longer fiber length (Koay et al. 2018a 2018b). Similarly, composite with higher fiber loading or fiber concentration would also enhance the composite properties. The highest concentration of fiber used was 60%, of the composite which study performed by Koay et al. (2018a 2018b) with durian husk fiber as the filler. Improvement of the MOE from 2700 MPa for pure PS to 2750 MPa for the composite and further increased to 3100 MPa with Ultra-Plast WP516 aid. This is due to when the amount of fiber loading increases, the greater the dispersion of filler particles throughout the matrix. Thus, it would increase the filler interaction with the PS matrix and reduce the gap between two of the composite materials (Sapuan & Bachtiar 2012).

In addition, pre-treatment of fiber using alkaline solution would enhance the composite mechanical properties by 20%. The result shown by Dixit and Yadav (2020) observed that pre-treating the agricultural waste fiber with sodium hydroxide as alkaline solution resulted in improved TS and thermogravimetric (Tg) value which are 3.56 MPa and 160 °C respectively compared to 2.3 MPa and 120 °C respectively for composite with untreated fiber. While comparing to pre-treating the agricultural waste fiber with hydrochloric acid, the TS and Tg values were much lower, which are 3.14 MPa and 150 °C, respectively (Dixit & Yadav 2020). Alkaline pre-treatment would remove lignin and hemicellulose from the surface of the fibers and increase the surface roughness of the fibers. This would improve the interfacial interlock between the treated fibers surface with the PS matrix, thus increase tensile and flexural strength of the composite material. Hemicellulose and lignin started to decompose at a much lower temperature around 250 °C as compared to cellulose as the major components in fiber which started to decompose at 350 °C, (Koay et al. 2018a 2018b). Thus, by removing hemicellulose and lignin, thermogravimetric value of the composite material can be improved.

Nevertheless, considering the usage of coupling agent, it would enhance the mechanical properties of the composite by 13%. This is due to the presence of coupling agent would stiffen the thin cell wall of agricultural fibers. Hence, it would sufficiently prevent the deformation of composite material under the tensile stress (Ghorbani et al. 2018). The addition of coupling agent also would improve the surface adhesion between the polystyrene matrix and fiber filler. For instance, highly reactive carbonyl groups (C=O) in maleic anhydride (MAPS) which is used as coupling agent are capable of forming ester bonds ($C–OCH_3$) with hydroxyl (–OH) groups of the fiber filler (Ling et al. 2019). Hence, the fiber surface becomes hydrophobic and increase the composite compatibility with hydrophobic polystyrene surface through non-polar covalent bond. Silane coupling agent also could be used as it would provide long-chain organic compound to attach to the fibers and make the fiber surface more hydrophobic (Shih et al. 2020).

4.3 *Pre-Treatment of Agricultural Waste for Polystyrene Reinforcement*

The limitation of polystyrene-agricultural waste composite is the compatibility of both materials to each other since agricultural waste is hydrophilic while polystyrene is hydrophobic. Treatment of agricultural waste is crucial to improve the composite properties since the purpose of PS composited with agricultural waste is to increase its functionality, energy density and hydrophobicity, creates porosity and making the biomass more accessible for cellulose region (Dixit & Yadav 2020). Hence the modification surface of agricultural waste to remove its hydrophilic chemical group such as hydroxyl (OH) group is required to increase its adherence to hydrophobic surface of polystyrene. The pre-treatment of agricultural waste can be performed by alkaline pre-treatment using sodium hydroxide (NaOH) solution to remove mainly lignin and hemicelluloses on the fiber surface (Stamatelatou et al. 2014). Furthermore, the modification of agricultural waste also can be done through esterification. It is an acylation procedure using strong acid such as hydrochloric acid (HCl), carboxylic acids as acylating agents by removing hydroxyl (–OH) from fiber surface. These two techniques are pre-treatment of agricultural fiber to reduce its hydrophilic properties and to achieve strong adhesion with polystyrene. Other less common approaches that can be used for the pre-treatment of the fiber include converting the agricultural waste to char through pyrolysis process to remove active hydrophilic group to make it compatible with PS (Attia et al. 2015; Adeniyi et al. 2020), modification of the cellulose fiber with TEMPO radical oxidized method or sol gel method.

4.4 *Pre-Treatment of Polystyrene for Composite Preparation*

Other than pre-treatment of agricultural waste, PS can also be modified to have compatible properties with agricultural waste fiber by the combination of coupling agents such as maleic anhydride with the processing aid such as Ultra-Plast WP516, a fatty acid amide compound. Coupling agent is added to improve the fiber-matrix adhesion while processing aid helps improve filler dispersion and processability of the composite (Ling et al. 2019). Other alternative may include using UV/ozone system (Yusilawati). UV photons will oxidize the PS surface involving the cleavage of unsaturated bond of ethylene which is present on the surface of PS to form free radicals then reacting with ozone to form oxidized species. Ozone will react with carbon surface to form ozonide. The degradation of ozonide by the presence of more ozone or oxygen as oxidizing agents will lead to the formation of hydroxyls, carboxyl groups, or carboxylic acids group (Yusilawati et al. 2010; Chamas et al. 2020). Thus, this will make polystyrene more hydrophilic and compatible to agricultural fiber which can be used as the reinforcement material. The other example of polystyrene modification methods includes sulfonation, a method which used concentrated sulfuric acid. However, concentrated sulfuric acid is highly corrosive and potentially explosive in concentrated form. To date, both of these alternative methods have not been employed to prepare the agricultural waste-polystyrene composite.

4.5 Properties of the Agricultural Waste Reinforced Polystyrene Composite

Composites tend to have the following characteristics; high strength, high modulus, low density, excellent resistance to rupture and fatigue, and low coefficient of thermal expansion. Table 5 mainly focused on the mechanical and thermal properties of the composite material from the most recent studies. Mechanical properties include the tensile strength (TS) properties, tensile modulus (MOE) properties, and flexural strength (MOR) properties. Meanwhile, thermal properties include studies that analyzed the thermogravimetric (Tg) value of the agricultural waste reinforced polystyrene composite. According to Table 5, PS composite was increased in tensile strength with the highest increment of 71% when reinforced with small diameter (250–355um) and higher concentration (15%) of alkaline-treated wheat husk. The composite material also was increased in tensile modulus with the highest increment of 93% when reinforced with alkaline-treated rice husk at 40% fiber concentration and increased in flexural strength with the highest increment of 45% compared to pristine PS when reinforced with a small diameter (250–355um) and higher concentration (15%) of wheat husk. As described earlier in Sect. 1.4.2; the tensile strength, tensile modulus, and flexural strength properties of composite material were affected by the fiber characteristics and concentration. However, in average, the composite material was reduced in thermogravimetric (Tg) value compared to the pristine PS with the highest decrement of 45% when PS was reinforced with agricultural waste such as wheat husk. Although Tg value of fiber is higher than the polystyrene, Tg of composite materials was lower than pure polystyrene due to the gradual evaporation of absorbed moisture from the fibers and the decomposition properties of the fibers itself. The Tg value could be improved by considering fire retardant such as magnesium hydroxide and di-ammonium hydrogen phosphate as well as incorporation of MWCNTs during the composition process (Attia et al. 2015; Eskander et al. 2018). Alternatively, prior to the composite process, the drying process of agricultural fibers is crucial to have a good thermogravimetric property of the composite material. In some of the studies, properties of pure PS values were not measured so proper comparison with the composite properties could not be made. Other properties that were observed in the studies but not discussed in this chapter include morphology and water absorption of the composite.

5 Conclusion and Future Trends

Since polystyrene is a non-degradable plastic material, it is good to composite with other material to improve the material's life cycle and reduce cradle to grave material. Based on these databases, it has been proven that the agricultural waste fiber is a potential reinforced material to prepare the polystyrene composite. Agricultural waste is a good option to composite with polystyrene since it has high mechanical strength, on the same time it is an eco-friendly material. Furthermore, both materials are abundant wastes worldwide. From the result analysis, the common agricultural wastes that have been used as reinforcement material for polystyrene include rice husk and wheat husk fiber. They were mostly prepared using the method of alkaline pre-treatment of fiber followed by homogenous mixing with melted polystyrene by the addition of coupling agent at a high temperature which is above 100 °C. The agricultural waste reinforced polystyrene composite was reported to improve in tensile strength, tensile modulus, and flexural strength compared to polystyrene alone. Although it would be only slightly reduced in thermal properties, it is still a comparable composite material could be applied in many applications. On the other note, polystyrene could be pre-treated with UV/ozone system to turn it into hydrophilic properties (Nor et al., 2011) which would increase its surface interaction with the agricultural waste thus simplify the methodology and minimize the usage of chemical during composite processing in future.

References

Abdulkareem, S., Amosa, M., et al. (2019). Development of natural fibre reinforced polystyrene (NFRP) composites: Impact resistance study. *IOP Conference Series: Materials Science and Engineering, 640*, 012059.

Adeniyi, A., Abdulkareem, S., & et al. (2020). Morphological and thermal properties of polystyrene composite reinforced with biochar from elephant grass (*Pennisetum purpureum*). *Journal of Thermoplastic Composite Materials* 1–16.

Akinyemi, B. A., Okonkwo, C. E., Alhassan, E. A., et al. (2019). Durability and strength properties of particle boards from polystyrene–wood wastes. *Journal of Material Cycles and Waste Management, 21*, 1541–1549. https://doi.org/10.1007/s10163-019-00905-6

Alyamaç, K. E., Baalbaki, O., et al. (2018). List of contributors. In R. Siddique & P. Cachim (Eds.), *Waste and supplementary cementitious materials in concrete* (pp. xi–xii). Woodhead Publishing.

Attia, N. F., Afifi, H. A., et al. (2015). Synergistic study of carbon nanotubes, rice husk ash and flame retardant materials on the flammability of polystyrene nanocomposites. *Materials Today: Proceedings, 2*(7), 3998–4005.

Binhussain, M. A., & El-Tonsy, M. M. (2013). Palm leave and plastic waste wood composite for out-door structures. *Construction and Building Materials, 47*, 1431–1435.

Brigham, C. (2018). Chapter 3.22—biopolymers: Biodegradable alternatives to traditional plastics. In B. Török & T. Dransfield (Eds.), *Green chemistry* (pp. 753–770), Elsevier.

Chamas, A., Moon, H., Zheng, J., Qiu, Y., Tabassum, T., Jang, J. H., & Suh, S. (2020). Degradation rates of plastics in the environment. *ACS Sustainable Chemistry & Engineering, 8*(9), 3494–3511. https://doi.org/10.1021/acssuschemeng.9b06635

Chun, K., Muhammad, N., Fahamy, Y., Yeng, C., Choo, H., Pang, M., & Tshai, K. (2018). Wood plastic Composites made from corn husk fiber and recycled polystyrene foam. *Journal of Engineering Science and Technology, 13*(11), 3445–3456. Retrieved from http://jestec.taylors.edu.my/Vol%2013%20issue%2011%20November%202018/13_11_01.pdf

Collier County. (2020). *The facts on Styrofoam: Reduce and reuse.* Retrieved December 13, 2020, from https://www.colliercountyfl.gov/your-government/divisions-s-z/solid-hazardous-waste-management/keeping-green-helpful-information-page/the-facts-on-styrofoam-reduce-and-reuse

Dixit, S., & Yadav, V. L. (2020). Comparative study of polystyrene/chemically modified wheat straw composite for green packaging application. *Polymer Bulletin, 77*, 1307–1326. https://doi.org/10.1007/s00289-019-02804-0

El-Zawawy, W. K., Ibrahim, M. M., et al. (2011). Characterization of the effects of lignin and lignin complex particles as filler on a polystyrene film. *Materials Chemistry and Physics, 131*(1), 348–357.

Eskander, S. B., Tawfik, M. E., & Tawfic, M. L. (2018). Mechanical, flammability and thermal degradation characteristics of rice straw fiber-recycled polystyrene foam hard wood composites incorporating fire retardants. *Journal of Thermal Analysis and Calorimetry, 132*, 1115–1124. https://doi.org/10.1007/s10973-018-6984-6

Falca, G., Musteata, V.-E., Behzad, A. R., Chisca, S., & Nunes, S. P. (2019). Cellulose hollow fibers for organic resistant nanofiltration. *Journal of Membrane Science, 586*, 151–161. https://doi.org/10.1016/j.memsci.2019.05.009

Garside, M. (2020). Global production capacity of polystyrene 2018 & 2024. *Statista. Chemicals & Resources: Chemical Industry.* Retrieved from https://www.statista.com/statistics/1065889/global-polystyrene-production-capacity/

Ghorbani, M., Biparva, P., & Hosseinzadeh, S. (2018). Effect of colloidal silica nanoparticles extracted from agricultural waste on physical, mechanical and antifungal properties of wood polymer composite. *European Journal of Wood and Wood Products, 76*, 749–757. https://doi.org/10.1007/s00107-017-1157-z

Gutiérrez, T. J., & Alvarez, V. A. (2017). Properties of native and oxidized corn starch/polystyrene blends under conditions of reactive extrusion using zinc octanoate as a catalyst. *Reactive and Functional Polymers, 112*, 33–44.

Ismail, M., Yassene, A., et al. (2011). Mechanical properties of rice straw fiber-reinforced polymer composites. *Fibers and Polymers, 12*, 648–656.

Jahirul, M. I., Rasul, M., et al. (2012). Biofuels production through biomass pyrolysis—a technological review. *Energies, 5*, 4952–5001.

Jasmin, A. F., & Kin, W. E. (2019). Plastic: An undegradable problem. *Khazanah Research Institute.* Retrieved from http://www.krinstitute.org/Views-@-Plastic-;_An_Undegradable_Problem.aspx

Jayamani, E., Rahman, M., Benhur, D., Bakri, M. K., Kakar, A., & Khan, A. (2020). Comparative study of fly ash/sugarcane fiber reinforced polymer composites properties. *BioResources, 15*, 5514–5531. https://doi.org/10.15376/biores.15.3.5514-5531

Khatib, J., & Negim, E.-S. (2013). Lightweight concrete made from waste polystyrene and fly ash. *World Applied Sciences Journal, 21*, 1356–1360. https://doi.org/10.5829/idosi.wasj.2013.21.9.20213

Koay, S. C., Fahamy, N. U. R., et al. (2018a). Wood plastic composites made from corn husk fiber and recycled polystyrene foam. *Journal of Engineering Science and Technology, 13*, 3445–3456.

Koay, S. C., Subramaniam, V., et al. (2018b). Wood plastic composites made from post-used polystyrene foam and agricultural waste. *Journal of Thermoplastic Composite Materials, 32*, 089270571879983.

Kumar, A., Negi, Y. S., Choudhary, V., & Bhardwaj, N. (2014). Sugarcane bagasse: A promising source for the production of nano-cellulose. *Journal of Polymer and Composites, 2*, 23–27.

Ling, S., Koay, S. C., et al. (2019). Wood plastic composites produced from postconsumer recycled polystyrene and coconut shell: Effect of coupling agent and processing aid on tensile, thermal, and morphological properties. *Polymer Engineering & Science*, 60.
Little, M. (2021). Facts about landfill & Styrofoam. *sciencing.com*. Retrieved from https://sciencing.com/facts-about-landfill-styrofoam-5176735.html. 6 July 2021.
Masri, T., Ounis, H., et al. (2018). Characterization of new composite material based on date palm leaflets and expanded polystyrene wastes. *Construction and Building Materials, 164*, 410–418.
MatWeb. (2020a). Overview of materials for Expanded Polystyrene (EPS). Retrieved November 30, 2020, from http://www.matweb.com/search/datasheet.aspx?matguid=5f099f2b5eeb41cba804ca0bc64fa62f
MatWeb. (2020b). Overview of materials for Polystyrene, Extrusion Grade. Retrieved June 30, 2021, from http://www.matweb.com/search/datasheettext.aspx?matguid=1c41e50c2e324e00b0c4e419ca780304
Neves, R. M., Lopes, K. S., Zimmermann, M. V. G., et al. (2019). Characterization of polystyrene nanocomposites and expanded nanocomposites reinforced with cellulose nanofibers and nanocrystals. *Cellulose, 26*, 4417–4429. https://doi.org/10.1007/s10570-019-02392-2
Pao, C. N., & Chan, M. Y. (2018). Properties and characterization of wood plastic composites made from agro-waste materials and post-used expanded polyester foam. *Journal of Thermoplastic Composite Materials, 32*, 089270571877287.
Pottmaier, D., Costa, M., et al. (2013). Comparison of rice husk and wheat straw: From slow and fast pyrolysis to char combustion. *Energy & Fuels, 27*(11), 71157125.
PubChem. (2022). PubChem. Retrieved September 14, 2022, from @pubchem website: https://pubchem.ncbi.nlm.nih.gov/#query=The%20chemical%20structure%20of%20PS
Rahman, A., Ulven, C., et al. (2017). Selection, pretreatment, and use of wheat bran for making thermoplastic composite.
Sapuan, S. M., & Bachtiar, D. (2012). Mechanical properties of sugar palm fibre reinforced high impact polystyrene composites. *Procedia Chemistry, 4*, 101–106.
Shih, Y. F., Kotharangannagari, V. K., & Tsou, T. C. (2020). Development of eco-friendly modified cellulose nanofiber reinforced polystyrene nanocomposites: Thermal, mechanical, and optical properties. *Journal of Polymer Research, 27*, 181. https://doi.org/10.1007/s10965-020-02156-8
Shuaib-Babata, Y. L., Busari, Y. O., Abdulraman, S. O., Abdulqadir, L. B., Ambali, I. O., Ajao, K. S., & Arowolo, I. O. (2022). The suitability of using environmental waste (polystyrene foam, sawdust and eggshells) for developing a hybrid composite for ceiling boards. *Nigerian Journal of Technology, 41*(1), 65–76. https://doi.org/10.4314/njt.v41i1.10
SpecialChem. (2020). Glass transition temperature.
Stamatelatou, K., Antonopoulou, G., & Michailides, P. (2014). 15—Biomethane and biohydrogen production via anaerobic digestion/fermentation. *Advances in Biorefineries, Woodhead Publishing* 476–524.
Tahir, T. A. (2012). *14 Chapter 1 introduction 1.1 waste composition in Malaysia*. Retrieved from http://studentsrepo.um.edu.my/3868/2/CHAPTER_1_%26_2_INTRODUCTION_%26_LIT_REVIEW.pdf
Wan Basri, W. N. F., Daud, H., Lam, M. K., Cheng, C. K., Oh, W. D., Tan, W. N., Kusakabe, K. (2019). A sugarcane-bagasse-based adsorbent employed for mitigating eutrophication threats and producing biodiesel simultaneously. *Processes, 7*(9), 572.
Yusilawati, A. N., Maizirwan, M., Hamzah, M. S., Ng, K. H., & Wong, C. S. (2010). Surface modification of polystyrene beads by ultraviolet/ozone treatment and its effect on gelatin coating. *American Journal of Applied Sciences, 7*, 724–731.
Zafar M. F., & Siddiqui M. A. (2019) Effect of filler parameters on the mechanical properties of wheat husk filled polystyrene composite. In I. Singh, P. Bajpai, & K. Panwar (Eds.), *Trends in materials engineering. Lecture notes on multidisciplinary industrial engineering*. Springer. https://doi.org/10.1007/978-981-13-9016-6_12

Zafar, M., & Siddiqui, M. (2020). Effect of filler loading and size on the mechanical and morphological behaviour of natural fibre-reinforced polystyrene composites. *Advances in Materials and Processing Technologies* 1–13.

Zafar, M., & Siddiqui, M. (2020). Preparation and characterization of natural fiber filled polystyrene composite using in situ polymerisation technique. *Advances in Materials and Processing Technologies* 1–11.

Zizumbo, A., Licea-Claveríe, A., Lugo-Medina, E., Garcia-Hernandez, E., Peralta, D., & Zitzumbo, R. (2011). Polystyrene composites prepared with polystyrene grafted fibers of sugarcane bagasse as reinforcing material. *Journal of the Mexican Chemical Society, 55*, 33–41.

Application of Fish Gelatine Nanoparticles in Food and Pharmaceutical Products

Irwandi Jaswir, Deni Subara, Widya Lestari, and Ridar Hendri

1 Introduction

Gelatin is a well-known protein that is derived from denatured collagen, which is processed from different parts of the animal skin, bone, and connective tissue. The process of collagen denaturation into gelatin involves inter-chain covalent crosslinking and some unfavorable breakage of the inter-chain peptide linkage (Zhou et al. 2006) that leaves gelatin with a lower molecular weight than native collagen consisting of hydrophobic fragments with molecular weight in the range of 50–250 kDa (Poppe 1997).

Gelatin has been used since ancient Egypt as wood glue (Schrieber et al. 2007). Until now, various applications of gelatin have been introduced which include coating agents, gelling, and cosmetic and pharmaceutical applications (Karim & Bhat 2008). In the pharmaceutical industry, gelatin is used to produce hard and soft capsules, and also to develop scaffolds for tissue generation (Karim & Bhat 2008). Around 1978, preparation of gelatin nanoparticles was introduced by Marty et al. (1978). Since then, studies on gelatin nanoparticles have increased.

I. Jaswir (✉)
International Halal Centre, Padang State University, Padang 25173, Indonesia
e-mail: irwandi@iium.edu.my

International Institute for Halal Research and Training (INHART), International Islamic University Malaysia, Jalan Gombak, 53100 Kuala Lumpur, Malaysia

D. Subara
Agroindustrial Technology Department, Institut Teknologi Sumatera, Lampung, Indonesia

W. Lestari
Kulliyyah of Dentistry, International Islamic University, Bandar Indera Mahkota, 25200 Kuantan, Malaysia

R. Hendri
Faculty of Fisheries and Marine Sciences, Riau University, Pekanbaru 28293, Indonesia

F. Adyani Ahmad Fuad et al. (eds.), *Circular Bioeconomy*,
https://doi.org/10.1007/978-981-97-7010-6_11

Nanoparticles are defined as colloidal system particles with sizes ranging from 1 to 1000 nm (Harmia et al. 1986; Sahoo et al. 2015). Nanoparticles are commonly investigated as a drug delivery strategy in pharmaceutical research. Nanoparticles were developed due to the challenges faced in drug delivery such as poor accuracy, safety and efficiency. Those challenges have driven researchers to actively search and publish in nanotechnology research, as pioneered by Speiser and co-workers (Marty et al. 1978; Sahoo et al. 2015). Since then, research in nanoparticles as a drug delivery system has increased. The use of nanoparticles to deliver drugs is based on several advantages, including the ability of nanoparticles to deliver the drug into different organs, e.g., the brain, lung, skin, and also tumors cells (Elzoghby 2013).

Nanoparticles are capable of carrying various drugs including proteins, vaccines, and nutraceuticals. The ability of nanoparticles to carry specific drugs can reduce unwanted side effects on target cells. Moreover, nanoparticles protect the drug from degradation, consequently increasing the bioavailability of the drug. This characteristic can decrease the required drug dosage. In addition, nanoparticles are capable of controlled and sustained release of drug. All of these advantages show the potential of nanoparticles and collectively describe why research in nanoparticles continues to gain interest.

2 Material for Nanoparticle Production

Nanoparticles can be prepared from different starting materials. Each material has different advantages and disadvantages, but they should comply with the requirements of biocompatibility and biodegradability. The starting material can be divided into two types, namely: synthetic polymers such as polyethylenimine (PEI), poly(lactic-co-glycolic)acid (PLGA), and natural polymers such as polysaccharides and lipids (Kreuter 2007; Liu et al. 2008; Weber et al. 2014).

Polyethylenimine or PEI has been used as a delivery system for non-viral genes, where the nucleic-acid of non-viral genes is attached to the cationic polymers by electrostatic interaction (Gharwan et al. 2003; Grayson et al. 2006). However, the limitation of PEI appears to be the lack of biodegradability and rather high toxicity of the material (Moghimi et al. 2005; Wen et al. 2009). Another starting material is PLGA. PLGA is a synthetic material that is widely used as a starting material to produce nanoparticles. It was approved by the FDA as GRAS due to its biocompatible and biodegradable properties. The particle size of PLGA can be controlled by controlling production variables. Due to the manufacturer's ability to control the final particle size, PLGA has been used to carry drugs for delivery to different parts of the immune system. However, complications arise when PLGA degrades, whereby the acidic components inside the particle are also released. This immediately drops the pH of the surrounding environment, which could affect the activity of the substances carried, such as proteins or nucleic acids.

Besides PEI and PLGA, lipids are also used as starting material for producing nanoparticles as they are also non-toxic and biodegradable. Solid lipid nanoparticles

(SLN) and liposomes are included under this type of material. SLNs were demonstrated to have great physical stability and physiological tolerability, and were able to protect the drug from degradation. However, SLNs were not able to control drug release due to their limited stability. During storage, the lipids undergo polymorphic transition. Lastly, liposome allows the researcher to incorporate various types of antigens. The structure of liposome is unique, consisting of an aqueous core surrounded by layers of lipids. This structure composition is quite similar to the cell membrane structure, which improves liposome compatibility with the cells. The liposome also protects the loaded drug from immediate degradation from the environment. However, the limitations of using liposomes are the inability to control drug distribution in the nanoparticle and subsequent removal in vivo.

3 Proteins as Nanoparticle Base Material

Between the different starting materials available for preparing nanoparticles, proteins have garnered the most interest among researchers. Some of the reported benefits of using protein as starting material for nanoparticle production include: biodegradability, biocompatibility, ease of size control, and amenability to surface modification for drug attachment (Elzoghby et al. 2011 2012). Protein-based nanoparticles have been recognized as generally regarded as safe (GRAS) by the FDA, with high natural nutritional value. Protein is readily available in large amounts; this allows it to be used as a less expensive starting material. It has also been reported that protein nanoparticles are better able to encapsulate poorly soluble drugs and increase their solubility. As an example, collagen nanoparticles are able to increase the delivery of poorly soluble transdermal drug through the skin compared to the freshly prepared drug (Lee et al. 2001). In addition, protein nanoparticles are able to control the release of the carried drug.

Despite the advantages of protein nanoparticles, there still are challenges that need to be solved. Natural protein consists of different molecular weight molecules in their composition; this makes the final product poor in uniformity and differ between batch-to-batch productions (Malafaya et al. 2007). Another limitation of protein nanoparticles is the risk of contamination with pathogens that can be present in mammalian proteins.

Generally, proteins for nanoparticle production can be classified into plant-based proteins and animal-based proteins. Plant-based proteins include zein, gliadin, and soy protein, whereas animal proteins include albumin, casein, silk protein, collagen, and gelatin. The advantage of using plant-based proteins compared to animal proteins is the hydrophobic character of plant-based proteins that remove the need for toxic cross-linkers in the production process (Ezpeleta et al. 1996). Additionally, plant-based proteins are cheaper than animal proteins (Ezpeleta et al. 1996).

Animal proteins are less advantageous than synthetic proteins for nanoparticle production due to their high potential toxicity (Leo et al. 1997). The major challenge in using animal proteins would be the risk of adulteration with pathogens, for example bovine spongiform encephalopathy (BSE), especially with mammalian gelatin. Hence, due to its similar composition, fish gelatin could be an attractive alternative for mammalian gelatin.

Numerous studies have been published on the impact of various processing factors on the properties of gelatin nanoparticles. These studies can be grouped into three categories: effect of processing variables, encapsulation of different drugs, and surface modification of gelatin nanoparticles. However, no report has been released regarding the production of fish gelatin nanoparticles, where almost all these studies used mammalian gelatin (either bovine or porcine gelatin) in their studies.

4 Collagen Versus Gelatin as Base Material for Nanoparticles

Despite gelatin and collagen both being built from amino acid chains, collagen is found to be not suitable for producing nanoparticles. Very few collagen-based drug delivery products have gone into clinical trials (Friess 1998). The three major reasons for this are: first, the high cost of producing pure collagen; second, the presence of large variations of impurities, crosslinking density, and fiber size in collagen, and third, complex handling properties of collagen (Friess 1998; Gan et al. 2011).

Besides, several reasons make gelatin more popular for use in nanoparticles compared to collagen: First, the gelatin matrix is better able to encapsulate valuable labile drugs such as vitamins, probiotics, and antioxidants. This is because gelatin matrices are built by various amino acids, including hydrophilic and hydrophobic amino acids. Secondly, gelatin is able to load larger amounts of drug. This is due to the high protein content in gelatin, which is able to link with the drug. The denaturation of collagen into gelatin under controlled pH and temperature conditions results in gelatin with high amino acid content. The amino acid consists of a repetitive glycine-X–Y sequence, whereas the imino acids Proline (Pro) and Hydroxyproline (Hyp) most frequently occupy the X–Y position (Li et al. 1997). Under wet conditions, the X and Y residues of gelatin build a triple-helix structure that is important for binding other molecules (Li et al. 1997). Thirdly, the high protein content in the gelatin structure makes gelatin a biocompatible and biodegradable material. This characteristic is due to gelatin being produced from natural sources. Biodegradable materials have many advantages, for example, they avoid rejection of the drug from the matrix of the organ after it is released (Li et al. 1997). Lastly, gelatin is a low-cost polymer and is abundant. Gelatin is produced from the skin, bone, and connective tissue of animals, which is readily available in most meat processing plants. Gelatin type A is produced from the acid processing of pig skin, and type B is produced from the basic processing of bovine bone (Sahoo et al. 2015).

5 Mammalian Gelatin and Fish Gelatin

Mammalian gelatin is the most readily available type of gelatin in the market. However, gelatin is also produced from non-mammalian sources, including chicken and fish skin and bone (Nagai & Suzuki 2000). The outbreak of Bovine Spongiform Encephalopathy (BSE) in Europe has increased the number of consumers avoiding mammalian-based products and switching to fish-based products. The demand for fish-based products is also made by religious groups that avoid consuming mammalian gelatin, such as Jews and Muslims, who are forbidden from consuming pig-based products, and Hindus, who are forbidden from consuming cow-related products. In this regard, rather than using mammalian gelatin as starting material, replacement with fish gelatin-based products is an appropriate alternative.

Fish gelatin is derived from fish by-products; it has the same composition as mammalian gelatin, but contains different concentrations of imino acids. Fish gelatin has lower concentrations of imino acids (proline and hydroxyproline) compared to mammalian gelatin (Eastoe & Leach 1977). Fish gelatin consists of approximately 22–25% imino acids, while mammalian gelatin consists of 30% imino acids (Muyonga et al. 2004). The low imino acid content decreases the gel modulus of fish gelatin, which lowers its gelling and melting temperature. This is because fish gelatin gels undergo the transformation of the super-helix structure, which is critical for the gel properties (Haug et al. 2004). Table 2.1 shows the amino acid composition of fish gelatin and mammalian gelatin.

Both mammalian as well as fish gelatin are used in many applications. Fish gelatin is used in food products, foam stabilization, low-fat spreads (provides creaminess, fat reduction, and mouth feel), dairy (provides stabilization and texture), baked goods (gelling, emulsion, and stabilization), and in meat products (provides water binding) (Frank 1990; Schrieber et al. 2007). In the pharmaceutical and medical industries, fish gelatin is used as a matrix implant, in injectable drug delivery microspheres and in intravenous infusion (Pollack 1990; Rao 1995; Saddler & Horsey 1987). However, no literature has been published on the application of fish gelatin to produce nanoparticles. Table 1 shows amino acid profile of fish and mammalian gelatin.

6 Fabrication Technique for Production of Gelatin Nanoparticles

Various methods for the preparation of mammalian gelatin nanoparticles have been documented in literature. The choice of production method is important in order to achieve the desired particle properties. The different properties of fish and mammalian gelatin create the need for different approaches in the nanoparticle preparation method. Hence, extensive studies continue to be carried out to optimize the production of fish gelatin nanoparticles. In this sub-chapter, the various methods are

Table 1 Amino acid profile of fish and mammalian gelatin

Amino acid	Tilapia skin collagen (Tanga et al. 2015)	Tilapia skin gelatin (Sarabia et al. 2000)	Calf skin gelatin (Piez 1960)
Proline (Pro)	107	119	138
Hydroxyproline (Hyp)	57	79	94
Threonine (Thr)	28	24	18
Serine (Ser)	37	35	36
Glutamic Acid (Glu)	74	69	72
Glycine (Gly)	319	347	320
Alanine (Ala)	123	123	112
Cysteine (Cys)	7	0	0
Valine (Val)	26	15	20
Methionine (Met)	16	9	4
Isoleucine (Ile)	14	8	11
Leucine (Leu)	29	23	25
Tyrosine (Tyr)	5	2	3
Phenylalanine (Phe)	17	13	13
Lysine (Lys)	29	25	27
Histidine (His)	5	6	5
Arginine (Arg)	55	47	50
Aspartic acid (Asp)	53	48	45

discussed in detail based on the principles, as well as the final size of the nanoparticle. The advantages and disadvantages of each production method are presented.

7 Nanoprecipitation

Nanoprecipitation was first introduced by Fessi et al. in 1989 (Quintanar-Guerrero et al. 1998). Two miscible solvents are required in this method, in addition to other materials including cross-linkers and stabilizers. Two phases are used in this method: the solvent phase and non-solvent phase. The non-solvent phase containing polymers is dropped into the solvent phase and mixed under continuous stirring, followed by the addition of cross-linker. In this method, the protein is dissolvable in non-solvent phase and non-dissolvable in solvent phase. Ethanol, acetone, hexane, and methylene chloride are mostly applied as the solvent phase, while water is mostly used as the non-solvent phase. This method comprises four steps: the first step is creating a super-saturation of solution, the second step is nucleation of proteins, the third step is growing the particles, and the last step is coagulation. Super-saturation is the state whereby the ability of the non-solvent phase to dissolve protein is decreased after mixing into the solvent phase.

The nanoprecipitation method has many advantages: it is a rapid and simple method because no extra pre-treatment of the starting material is needed, there is no need to remove the low-molecular weight portion of gelatin, it produces satisfying reproducibility, the formulation produced has good stability, and there is no need for sonication or high temperatures. The limitation of using nanoprecipitation is the inability to control particle growth; consequently, this method needs high amounts of surfactants (Miladi et al. 2016).

In the nanoprecipitation technique, two solvents that are miscible with one another, a cross-linker, and a stabilizer are required. The principle of this technique is based on the forced diffusion of water molecules in gelatin to create small particles with different solubilities compared to their environment. Gelatin is dissolved in one solvent such as water and slowly dropped into the non-solvent phase such as ethanol or acetone. Then, glutaraldehyde is added to crosslink the nanoparticles in the presence of a surfactant such as poloxomer as stabilizer.

8 Simple Water/Oil Emulsion

Jia Kui Li et al., in 1998 introduced a simple emulsion method using gelatin as starting material and bovine serum albumin as a model protein drug (Li et al. 1998). Generally, the emulsion method consists of two immiscible phases of solvent: the first phase is the oil phase, and the non-solvent phase is the water solvent. This method consists of three preparation steps. In this method, the gelatin and drug are first diluted in water solvent. Then, the water solvent is mixed with the oil phase and surfactants by high-speed homogenization. The process is then continued by adding cross-linkers such as glutaraldehyde. The nanoparticles are then washed and extracted from the mixture. The most common oil solvents used in the nanoemulsion process are organic solutions of paraffin or polymethylmethacrylate.

This method is advantageous for producing nanoparticles that are mostly thermodynamically stable and is suitable for the delivery of lipophilic drugs (Singh et al. 2017). However, despite its advantages, there are several disadvantages of the emulsion method for particle preparation, such as the need for huge amounts of oil and organic solvent. This solution needs to be discarded as it may contaminate the final product (Rahimnejad et al. 2012).

9 Coacervation

In this technique, gelatin is dissolved in water as a non-solvent phase. The coacervation technique consists of a solvent phase, non-solvent phase, cross-linker, and stabilizer. The difference between coacervation and nanoprecipitation is in the coacervation method, the non-solvent phase is dropped into the solvent phase, while in the nanoprecipitation method, the solvent phase is dropped into the non-solvent phase. Another difference is in the coacervation method, nucleation and particle growth are created by the addition of natural salts because it promotes the hydration of the solute. This addition of salt leads to liquid–liquid phase separation whereby the dense particles are left in the bottom of the solution and the transparent fraction is left at the top. Basically, this technique involves dehydration of the protein molecules (Leo et al. 1997), nucleation of proteins, growth of the particles and lastly, coagulation.

The coacervation method offers several advantages; it does not need complex machinery, sonication, and high temperature environment (Hao et al. 2012). However, this technique has its disadvantages, such as strict pH requirements during the process, whereby coacervation only functions in a limited pH range (Comunian et al. 2013). In addition, this technique is not suitable for the encapsulation of salt-sensitive drugs and needs proper techniques for the complete removal of salts from the nanoparticle product (Mishra 2015).

10 Desolvation

Desolvation is a simple, widely applied method used to prepare gelatin nanoparticles (Azarmi et al. 2006; Karthikeyan et al. 2013; Petrochenko et al. 2012). This method consists of the solvent phase, non-solvent phase, and cross-linker, without employing stabilizers. Generally the desolvation method is based on the principle of dehydration of gelatin molecules due to the addition of solvent phase to the mixture. The solvent phase is dropped into an aqueous gelatin solution under vigorous stirring. Most of the time, the non-solvent phase is water, while acetone, ethanol, and methanol are used as the solvent phase. The advantages of the desolvation method are: no extended step needed, no sonication or high temperature application during the process, and no extra stabilizer added. However, the production of nanoparticles using this method is dependent on several aspects, such as pH during desolvation, purity of starting material and choice of cross-linker agent (Stevenson et al. 2018).

11 Two-Step Desolvation

The two-step desolvation method was introduced by Coester et al. (2000). Generally, this method was developed to solve the problem of impurities present in the starting material in the desolvation method. The trick involves adding a quick desolvation step at the beginning of the process, before drop-wise addition of solvent in the second desolvation. In the first desolvation step, low-molecular weight gelatin and high-molecular weight gelatin are separated. Previous studies report that the size of gelatin nanoparticles is affected by the molecular weight heterogeneity of the starting material (Stevenson et al. 2018). This method involves four steps: separating high-molecular weight gelatin, hydration of the gelatin molecule, nucleation of the gelatin molecule, and crosslinking. Although this method was created to improve on the desolvation method, two-step desolvation still faces challenges in the reproducibility of batch-to-batch production. Furthermore, the two-step desolvation method produces low yields of nanoparticles because only the high-molecular weight fraction of gelatin is used (Balthasar et al. 2005). Based on this challenge, studies on the effects of different production factors on the particle produced have been carried out.

The two-step desolvation method consists of two steps. In the first step, the main target is the separation of high-molecular weight gelatin from low-molecular weight gelatin, while in the second step, the main goal is to produce nanoparticles via drop-wise addition of acetone and crosslinking with cross-linker. Several studies have been published optimizing the two-step desolvation process. Some of these studies have focused on the first step, while others have focused on the second step to achieve small and uniform-sized nanoparticles (Table 2).

Table 2 Advantages and disadvantages of different production methods for gelatin nanoparticles

Preparation methods	Size	Positive aspects	Negative aspects	References
Two-step desolvation	100–300	Simple procedure, homogenous size	Narrow pH range, specific molecular weight requirement	Stevenson et al., (2018)
Desolvation	200–500	Simple procedure	Agglomeration, stability issues	Geh et al., (2016)
Simple W/O emulsion	100–200	Homogenous size	Long washing procedure	Tomar et al., (2013)
Nanoprecipitation	250–350	Simple, straight forward procedure	High amount of surfactant needed	Khan and Schneider (2013)

12 Food and Pharmaceutical Applications of Fish Gelatin Nanoparticles

12.1 Application of Fish Gelatin Nanoparticles in Food

Fish gelatin nanoparticles have been used to carry bioactive materials for food application. Iqrah Akbar (2020) studied the application of fish gelatin nanoparticles for the delivery of antioxidant biopeptides in yogurt. In his study, production of the antioxidant biopeptide-gelatin nanoparticles was designed using Response Surface Methodology (RSM). Critical independent variables that had significant influence on the response i.e., gelatin nanoparticle size and encapsulation efficiency (EE) were used, including pH, crosslinking agent 'glutaraldehyde' volume and biopeptide concentration. A total of 20 experimental runs using Central Composite Design were carried out to study the effect of the selected independent variables on the responses. Analysis of Variance showed that the predicted models were statistically valid.

The optimized condition for the production of antioxidant biopeptide-gelatin nanoparticles with desired properties was pH 2, 2 ml glutaraldehyde and 0.1 mg/ml antioxidant biopeptide. The nanoparticles produced had size 228.3 ± 0.11 nm, encapsulation efficiency 76 ± 1.1%, DPPH activity 92%, and FRAP assay values of 978 ± 0.65 μm/L. The experimental values obtained from the optimization steps were within the range of predicted values using RSM. Verification experiments conducted under optimal conditions to compare predicted values and actual values of responses showed similar results. Therefore, the estimated RSM was adopted for optimization of gelatin nanoparticle production encapsulating sunflower-derived antioxidant biopeptide.

A study on the gastro intestinal digestion of the gelatin nanoparticles loaded with antioxidant nanoparticles showed that the gelatin nanoparticles could easily be digested via the action of pepsin and pancreatin, the two enzymes responsible for protein digestion in the human stomach and small intestine. The protein released from gelatin nanoparticles increased to 0.1464 mg/mL after pepsin digestion, followed by 0.277 mg/mL after pancreatin digestion. This is consistent with the changes noticed in the gastro intestinal uptake of proteins, thereby suggesting that the protein release from the gelatin nanoparticles was satisfactory (Akbar 2020).

Cytotoxicity studies were performed on human fibroblasts using gelatin nanoparticles at various concentrations. Addition of gelatin nanoparticles did not affect the number of human fibroblast cells on glass compared to control cells containing no particles, and the cells maintained their morphology and adhesion capacity. Furthermore, MTT assay demonstrated that cells incubated with gelatin nanoparticles remained more than 100% viable at concentrations as high as 25 μg/ml, indicating low toxicity of gelatin nanoparticles in vitro in human fibroblasts and potentially low risk in vivo. The results suggest that encapsulation of the antioxidant biopeptide, followed by its release does not contain harmful by-products which can have a deleterious effect on the consumers of the gelatin nanoparticles added in food.

A comparative study on the physical, chemical, and antioxidant properties of yogurt was done to study the effect of the addition of gelatin nanoparticles loaded with the antioxidant biopeptide for nutritional fortification of yogurt. Physical analysis revealed no significant change in hardness or consistency upon the addition of gelatin nanoparticles; however, stickiness slightly changed. Comparison of the chemical properties of yoghurt with and without gelatin nanoparticles showed negligible change in terms of pH, but significant difference in syneresis of the fortified yoghurt. Furthermore, addition of the gelatin nanoparticles improved antioxidant activity of the yoghurt, where DPPH values increased from 79.0 to 83.4% and FRAP values improved from 869.0 to 978.0 ± 0.65 μm/L, concluding that yogurt fortified with the biopeptide-gelatin nanoparticles increased antioxidant activity to potentially fulfill consumer requirements.

12.2 Application of Fish Gelatin Nanoparticles in Pharmaceuticals

Fish gelatin nanoparticles may also be a promising drug carrier for pharmaceutical applications. Subara et al. (2018) reported that fish gelatin nanoparticles could be used to carry hydrophilic and hydrophobic drugs, such as salicylic acid and 5-fluorouracil, respectively. In order to encapsulate the drug into fish gelatin nanoparticles, significant factors were selected through screening using factorial design, and the selected factors were optimized afterward. Drug amount was the only significant factor for encapsulation efficiency, where pH, acetone percentage, glutaraldehyde amount, and stirring time were non-significant factors. Drug-loaded fish gelatin nanoparticles with an encapsulation efficiency of 39% were produced using 0.33 g salicylic acid.

Two-stage drug release from fish gelatin nanoparticles was observed, whereby a burst release of 50% was observed in the initial 5 h incubation. The drug release mechanism followed non-Fickian model, as the exponent n from the Korsmeyer-Peppas model was around 0.7. This burst release indicated that drug loading in fish gelatin nanoparticles was via adsorption on the nanoparticle surface and entrapment. These results reveal that the drug release pattern from fish gelatin nanoparticles may be applied in biological systems that require a rapid response. Based on SEM images, the unloaded nanoparticles and drug-loaded nanoparticles were spherical in shape and had a smooth surface. The AFM and TEM images showed that the nanoparticles were dispersed uniformly, consistent with SEM observations. FTIR indicated that physical interactions between the core and wall matrices, rather than chemical interactions, influenced encapsulation efficiency. Meanwhile, XRD showed that salicylic acid was amorphously dispersed within the nanoparticle matrix. DSC analysis revealed that the molecular dispersion of the drug was as a disordered crystalline phase in the nanoparticle matrix. TGA revealed that the drug encapsulation process slightly increased the thermal stability of the nanoparticles to a higher temperature range.

Acknowledgements The authors thank Lembaga Penelitian and Pengabdian Masyarakat (LPPM), Padang State University for funding this study through RKAT Fund 2024 with International Collaboration.

References

Akbar, I. (2020). *The development of tilapia fish skin gelatin nanoparticles in encapsulated biopeptides for yoghurt application*. International Islamic University Malaysia.

Azarmi, S., Huang, Y., Chen, H., Azarmia, S., Huang, Y., & Chen, H. (2006). Optimization of a two-step desolvation method for preparing gelatin nanoparticles and cell uptake studies in 143B osteosarcoma cancer cells. *Journal of Pharmaceutical Sciences, 9*(1), 124–132.

Balthasar, S., Michaelis, K., Dinauer, N., Briesen, H. von, Kreuter, J., & Langer, K. (2005). Preparation and characterisation of antibody modified gelatin nanoparticles as drug carrier system for uptake in lymphocytes. *Biomaterials, 26*, 2723–2732.

Coester, C., Langer, K., Von Briesen, H., & Kreuter, J. (2000). Gelatin nanoparticles by two step desolvation—a new preparation method, surface modifications and cell uptake. *Journal of Microencapsulation, 17*(2), 187–193.

Comunian, T. A., Thomazini, M., Julia, A., Alves, G., Eustáquio, F., Junior, D. M., & Favaro-trindade, C. S. (2013). Microencapsulation of ascorbic acid by complex coacervation: Protection and controlled release. *Food Research International, 52*(1), 373–379.

Eastoe, J. E., & Leach, A. A. (1977). Hemical constitution of gelatin. In A. G. Ward & A. Courts (Eds.), *The science and technology of gelatin* (pp. 73–107). Academic Press.

Elzoghby, A. O. (2013). Gelatin-based nanoparticles as drug and gene delivery systems: Reviewing three decades of research. *Journal of Controlled Release, 172*(3), 1075–1091.

Elzoghby, A. O., El-fotoh, W. S. A., & Elgindy, N. A. (2011). Casein-based formulations as promising controlled release drug delivery systems. *Journal of Controlled Release, 153*(3), 206–216.

Elzoghby, A. O., Samy, W. M., & Elgindy, N. A. (2012). Albumin-based nanoparticles as potential controlled release drug delivery systems. *Journal of Controlled Release, 157*(2), 168–182.

Ezpeleta, I., Irache, J. M., Stainmesse, S., Chabenat, C., Popineau, Y., & Orecchionp, A. (1996). Gliadin nanoparticles for the controlled release of all-trans-retinoic acid. *International Journal of Pharmaceutics, 131*(2), 191–200.

Frank, A. (1990). Gelatin. In *Food gels* (pp. 233–289). Elsevier Applied Sciences.

Friess, W. (1998). Collagen—biomaterial for drug delivery. *European Journal of Pharmaceutics and Biopharmaceutics, 45*, 113–136.

Gan, Z., Ju, J., Zhang, T., & Wu, D. (2011). Preparation of rhodamine B fluorescent poly(methacrylic acid) coated gelatin nanoparticles. *Journal of Nanomaterials, 2011*.

Geh, K. J., Hubert, M., & Winter, G. (2016). Optimization of one-step desolvation and scale-up of gelatin nanoparticle production. *Journal of Microencapsulation, 33*(7), 1–10.

Gharwan, H., Wightman, L., Kircheis, R., Wagner, E., & Zatloukal, K. (2003). Nonviral gene transfer into fetal mouse livers (a comparison between the cationic polymer PEI and naked DNA). *Gene Therapy, 10*, 810–817.

Grayson, A. C. R., Doody, A. M., & Putnam, D. (2006). Biophysical and structural characterization of polyethylenimine-mediated siRNA delivery in vitro. *Pharmaceutical Research, 23*(8), 1868–1876.

Hao, J., Wang, F., Wang, X., Zhang, D., Bi, Y., Gao, Y., & Zhang, Q. (2012). Development and optimization of baicalin-loaded solid lipid nanoparticles prepared by coacervation method using central composite design. *European Journal of Pharmaceutical Sciences, 47*(2), 497–505.

Harmia, T., Kreuter, J., Speiser, P., Boye, T., Gurny, R., & Kubi, A. (1986). Enhancement of the myotic response of rabbits with pilocarpine-loaded polybutylcyanoacrylate nanoparticles. *International Journal of Pharmaceutics, 33*(1–3), 187–193.

Haug, I. J., Draget, K. I., & Smidsrød, O. (2004). Physical and rheological properties of fish gelatin compared to mammalian gelatin. *Food Hydrocolloids, 18*, 203–213.

Karim, A. A., & Bhat, R. (2008). Gelatin alternatives for the food industry: Recent developments, challenges and prospects. *Trends in Food Science and Technology, 19*(12), 644–656.

Karthikeyan, S., Prasad, N., Ganamani, A., & Balamurugan, E. (2013). Anticancer activity of resveratrol-loaded gelatin nanoparticles on NCI-H460 non-small cell lung cancer cells. *Biomedicine & Preventive Nutrition, 3*, 64–73. Retrieved from https://doi.org/10.1016/j.bionut.2012.10.009

Khan, S. A., & Schneider, M. (2013). Nanoprecipitation versus two step desolvation technique for the preparation of gelatin nanoparticles. *Procedia of SPIE, 8595*, 85950H.

Kreuter, J. (2007). Nanoparticles-a historical perspective. *International Journal of Pharmaceutics, 331*(1), 1–10.

Lee, C. H., Singla, A., & Lee, Y. (2001). Biomedical applications of collagen. *International Journal of Pharmaceutics, 221*, 1–22.

Leo, E., Vandelli, M. A., Cameroni, R., & Forni, F. (1997). Doxorubicin-loaded gelatin nanoparticles stabilized by glutaraldehyde: Involvement of the drug in the cross-linking process. *International Journal of Pharmaceutics, 155*, 75–82.

Li, J. K., Wang, N., & Wu, X. S. (1998). Gelatin nanoencapsulation of protein/peptide drugs using an emulsifier-free emulsion method. *Journal of Microencapsulation, 15*, 163–172.

Li, J. K., Wang, N., & Wu, X. S. (1997). A novel biodegradable system based on gelatin nanoparticles and poly(lactic-co-glycolic acid) microspheres for protein and peptide drug delivery. *Journal of Pharmaceutical Sciences, 86*(8), 891–895.

Liu, Z., Jiao, Y., Wang, Y., Zhou, C., & Zhang, Z. (2008). Polysaccharides-based nanoparticles as drug delivery systems. *Advanced Drug Delivery Reviews, 60*(15), 1650–1662.

Malafaya, P. B., Silva, G. A., & Reis, R. L. (2007). Natural—origin polymers as carriers and scaffolds for biomolecules and cell delivery in tissue engineering applications. *Advanced Drug Delivery Reviews, 59*, 207–233.

Marty, J. J., Oppenheim, R. C., & Speiser, P. (1978). Nanoparticles--a new colloidal drug delivery system. *Pharmaceutica Acta Helvetiae, 53*(1), 17—23. Retrieved from http://europepmc.org/abstract/MED/643885

Miladi, K., Sfar, S., Fessi, H., & Elaissari, A. (2016). Nanoprecipitation process : From particle preparation to in vivo applications. In C. Vauthier & G. Ponchel (Eds.), *Polymer nanoparticles for nanomedicines* (pp. 17–53). Springer International Publishing Switzerland.

Mishra, M. (2015). *Handbook of encapsulation and controlled release.*

Moghimi, S. M., Symonds, P., Murray, J. C., Hunter, A. C., Debska, G., & Szewczyk, A. (2005). A two-stage poly (ethylenimine)-mediated cytotoxicity: Implications for gene transfer/therapy. *Molecular Therapy, 6*, 990–995.

Muyonga, J. H., Cole, C. G. B., & Duodu, K. G. (2004). Extraction and physico-chemical characterisation of Nile perch (*Lates niloticus*) skin and bone gelatin. *Food Hydrocolloids, 18*, 581–592.

Nagai, T., & Suzuki, N. (2000). Isolation of collagen from fish waste material-skin, bone and fins. *Food Chemistry, 68*, 277–281.

Petrochenko, P. E., Zhang, Q., Wang, H., Sun, T., Wildt, B., Betz, M. W., & Narayan, R. J. (2012). In vitro cytotoxicity of rare earth oxide nanoparticles for imaging applications. *International Journal of Applied Ceramic Technology, 9*(5), 881–892.

Piez, K. A. (1960). The amino acid composition of some fish collagens: The relation between composition and structure. *Gross, Jerome, 235*(4), 995–998.

Pollack, S. V. (1990). Silicone, fibrel, and collagen implantation for facial lines and wrinkles. *Journal of Dermatology and Surgical Oncology, 16*, 957–961.

Poppe, J. (1997). Gelatin. In *Thickening and gelling agents for food* (2nd edn, pp. 158–182). Springer.
Quintanar-Guerrero, D., Allémann, E., Fessi, H., & Doelker, E. (1998). Preparation techniques and mechanism of formation of biodegradable nanoparticles from preformed polymers. *Drug Development and Industrial Pharmacy, 24*(12), 1113–1128.
Rahimnejad, M., Najafpour, G., & Bakeri, G. (2012). Investigation and modeling effective parameters influencing the size of BSA protein nanoparticles as colloidal carrier. *Colloids and Surfaces a: Physicochemical and Engineering Aspects, 412*, 96–100.
Rao, K. P. (1995). Recent developments of collagenbased materials for medical applications and drug delivery systems. *Journal of Biomaterials Science, 7*, 623–645.
Saddler, J. M., & Horsey, P. J. (1987). The new generation gelatins. A review of their history, manufacture and properties. *Anesthesia, 42*, 998–1004.
Sahoo, N., Sahoo, R. K., Biswas, N., Guha, A., & Kuotsu, K. (2015). Recent advancement of gelatin nanoparticles in drug and vaccine delivery. *International Journal of Biological Macromolecules, 81*, 317–331.
Sarabia, A. A. I., Gómez-Guillén, M., Montero, P., Mez-GuilleÂn, M. C. G., & Montero, P. (2000). The effect of added salts on the viscoelastic properties of fish skin gelatin. *Food Chemistry, 70*, 71–6.
Schrieber, R., Herbert, G., & Gareis, H. (2007). *Gelatine handbook: Theory annd Industrial Practice*. Wiley-VCH GmbH & Co.
Singh, Y., Gopal, J., Raval, K., Ali, F., Chaurasia, M., Jain, N. K., & Chourasia, M. K. (2017). Nanoemulsion: Concepts, development and applications in drug delivery. *Journal of Controlled Release, 252*, 28–49.
Stevenson, A. T., Jankus, D. J., Tarshis, M. A., & Whittington, A. R. (2018). The correlation between gelatin macroscale differences and nanoparticle properties: Providing insight into biopolymer variability. *Nanoscale, 10*(21), 10094–10108.
Subara, D., Jaswir, I., Fahmi, M., Alkhatib, R., & Noorbatcha, I. A. (2018). Synthesis of fish gelatin nanoparticles and their application for the drug delivery based on response surface methodology. *Advances in Natural Sciences: Nanoscience and Nanotechnology, 9*, 1–11.
Tanga, L., Chena, S., Sua, W., Wenga, W., Osakoc, K., Tanaka, M., & Tanaka, M. (2015). Physicochemical properties and film-forming ability of fish skin collagen extracted from different freshwater species. *Process Biochemistry, 50*(1), 148–155.
Tomar, L., Tyagi, C., Kumar, M., Kumar, P., Singh, H., Choonara, Y. E., & Pillay, V. (2013). In vivo evaluation of a conjugated poly(lactide-ethylene glycol) nanoparticle depot formulation for prolonged insulin delivery in the diabetic rabbit model. *International Journal of Nanomedicine, 8*, 505–520.
Weber, S., Zimmer, A., & Pardeike, J. (2014). Solid lipid nanoparticles (SLN) and nanostructured lipid carriers (NLC) for pulmonary application: A review of the state of the art. *European Journal of Pharmaceutics and Biopharmaceutics, 86*(1), 7–22.
Wen, Y., Pan, S., Luo, X., Zhang, X., Zhang, W., & Feng, M. (2009). A biodegradable low molecular weight polyethylenimine derivative as low toxicity and efficient gene vector. *Bioconjugate Chem, 20*(2), 322–332.
Zhou, P., Mulvaney, S. J., & Regenstein, J. M. (2006). Properties of Alaska pollock skin gelatin: A comparison with Tilapia and pork skin gelatins. *Journal of Food Science, 71*(6), 313–321.

Plant Wastes to Biomaterials: Toward a Circular and Sustainable Economy

Nur Izzati Binti Mohd Razali, Maizatul Akmal Johari, Amina Tahreen, Mohammed Soleiman Barre, and Fathilah Binti Ali

1 Introduction

Agricultural sector is one of the three major sectors that contributes to Malaysian income. Department of statistics Malaysia official portal reported that in 2019, agriculture contributed 7.29% of gross domestic product (GDP). Among all agricultural activities, palm oil is still in the first place of major contributor in the agriculture sector at 37.7%. Besides, there are several others agricultural activities that contribute to the economic development such as the food production. All these agricultural activities produced large numbers of wastes and will contribute to bigger wastes problems if there is no proper management. In recent years, the awareness of recycling and reusing the agricultural wastes manages to turn the waste into many valuable products. To realize "waste to wealth" expansion, research, and development sector continue to discover several potentials of turning wastes into profit as well as focusing on targeting zero wastes. Agricultural wastes that can be found abundantly in Malaysia are included banana trunk, cassava, pineapple leaf, sago, sugarcane, and sweet potatoes. They have the potential to be used as a biomass that can produce other valuable byproducts which may involve in innovation of polymeric materials reinforcement. However, biopolymers must compete with polymers derived from fossil fuel in terms of functional properties where the petroleum-based plastics have remarkably higher tensile strength as well as hard to break when compared to biopolymer-based plastics. One of the downsides of petroleum-based plastics is they take longer time to degrade thus will contribute to both land and water pollution. To

N. I. B. M. Razali · M. A. Johari · A. Tahreen · M. S. Barre
Department of Biotechnology Engineering, Kulliyyah of Engineering, International Islamic University Malaysia, Jalan Gombak, 53100 Kuala Lumpur, Selangor, Malaysia

F. B. Ali (✉)
Department of Chemical Engineering and Sustainability, Kulliyyah of Engineering, International Islamic University Malaysia, Jalan Gombak, 53100 Kuala Lumpur, Selangor, Malaysia
e-mail: fathilah@iium.edu.my

F. Adyani Ahmad Fuad et al. (eds.), *Circular Bioeconomy*,
https://doi.org/10.1007/978-981-97-7010-6_12

achieve Sustainable Development Goals (SDGs), the use of nonrenewable resources should be controlled wisely and more exploration and exploitation on the renewable resources are needed.

2 Plant-Based Biopolymers

With advancements in the chemical and petrochemical production industries, remarkable economic boost has been harvested to meet global demands. However, there is a rising concern regarding the resulting environmental pollution and scarcity of resources, that requires strict consideration for the long-term wellbeing of the ecosystem. Everything in nature inspires a cyclic nature of events starting from the growth of a plant from soil to fruit harvesting and the wastes returned to the soil again, enriched with decomposed nutrients for further growth cycles of vegetation. The sustainability of nature maintained with its cyclic renewable patterns is the inspiration for adopting and implementing the principal in all aspects of life, particularly the circular economy. Hence, utilizing carbon rich plant wastes for useful eco-friendly biomaterials paves way for a sustainable circular economy incorporating the idea of turning waste to wealth.

Media (Ellen MacArthur Foundation 2017) repeatedly reported that polymers would need hundreds of years to be fully degraded therefore the usage of polymer has always been associated with a spike in global pollution rate. In the light of the concerns, biopolymers have become the best alternatives to those polymers since biopolymers are essentially biobased and biodegradable. Biopolymers are defined as polymers that are produced from natural sources, either naturally synthesized or chemically synthesized (Smith et al. 2016). Polylactic acid (PLA), polyhydroxyalkanoate (PHA), and carboxymethyl cellulose (CMC) are some of the examples of biopolymers that have been extensively studied. Among them, PLA is dominating the production of biopolymers, where NatureWork LLC is the leading producer of PLA with over 95% of productivity (Abdel-Rahman et al. 2013).

2.1 Polylactic Acid (PLA)

Chemically synthesized PLA produced a racemic of D- and L-lactic. This racemic increase the cost during the separation process. Conversely, PLA synthesized from microbial fermentation process can produce D- or L-lactic isomers, depending on the microorganism used during the process (Ahmad et al. 2020). Generally, PLA produced via fermentation pathways utilizes high carbon sources, therefore sugar cane molasses (Flores-Albino et al. 2012; Xu & Xu 2014), Jerusalem artichoke powder (Wang et al. 2013), apple pomace (Gullón et al. 2008), kitchen waste (Zhang et al. 2008), and paper slurry (Budhavaram & Fan 2009) are used as the feedstock in PLA fermentation process. Some of the above-mentioned feedstocks are from plant wastes therefore instead of increasing the cost to treat the waste, they can be used as feedstocks of another process. This kind of reusing waste into producing new materials is a part of the circular economy system (Ellen MacArthur Foundation 2017) and is expected to boost the economic growth of the related parties.

As mentioned previously, PLA can be synthesized either through chemical synthesis or fermentation process. In large-scale production, producing PLA through the fermentation process is preferrable. Microbial fermentation process is the most common protocol used. Bacteria, fungi, yeast, cyanobacteria, and microalgae are examples of microorganisms that able to produce lactic acid via fermentation process. Bacteria are commonly used in various fermentation process, including lactic acid fermentation process. Bacteria used in lactic acid fermentation can be categorized into 4 groups, known as lactic acid bacteria (LAB), *Bacillus* strain, *Escherichia coli*, and *Corynebacterium glutamicum*.

LAB consists of various group of Gram-positive microorganisms that exists within meat, plant, and dairy products. They can produce lactic acid as an anaerobic glycolysis product with high productivity and high yield. These bacteria can grow in the pH range of 3.5–10.0 and temperature range of 5–45 °C. LAB metabolized pentoses and hexoses via pentose phosphate pathways, glycolysis pathway, and phosphoketolase pathway (Abdel-Rahman et al. 2013). LAB that produced lactic acid as its major end-product and possess aldolase enzyme is known as homofermentative LAB whereas LAB that produced byproducts other than lactic acid is known as heterofermentative LAB. The former LAB is of interest for commercial scale lactic acid production while the latter only manage to produce lactic acid at 0.5 g/g or 1.0 mol/mol maximum yield (Abdel-Rahman et al. 2011). Most LAB, including *Lactobacilli*, are regarded to be safe for industrial lactic acid production because they have had a long history of industrial-scale production without adverse health effects on either consumers or production workers.

Commercially important LAB strains, such as *Lactobacillus* strains, have been particularly useful due to their high acid tolerance and their capability to be engineered for selective production of D-or L-lactic acid (Benthin & Villadsen 1995; Kyla-Nikkila et al. 2000; Lapierre et al. 1999). Alkaliphilic or thermotolerant LAB strains may be promising producers of lactic acid due to their tolerance to high pH levels that would minimize contamination problems during processing (Calabia et al. 2011; Liu et al. 2010). Calabia and team isolated an alkaliphilic *Halolactibacillus halophilus* from a marine environment that produced 65.8 g/L of L-lactic acid at pH 9.0 (Calabia et al. 2011). Although lactic acid production by LAB is very efficient, further improvements in the strain (especially, high acid tolerance) and the process can help to make it more cost-competitive with petroleum-based polymers for PLA production.

Bacillus species such as *Bacillus coagulans*, *Bacillus stearothermophilus*, *Bacillus licheniformis*, *Bacillus subtilis*, and *Bacillus* sp. have all been found to produce lactic acid. *Bacillus* spp., in comparison to LAB, have various potential advantages in lactic acid production that could assist in reducing lactic acid fermentation expenses, as follows: (i) *Bacillus* sp. can grow and produce lactic acid in mineral salt medium with few nitrogen sources rather than costly media (Wang et al. 2011); (ii) Meng et at. (2012) isolated alkaliphilic strains *Bacillus* sp. WL-S20, which produced L-lactic acid at a concentration of 225 g/L and a yield of 0.993 g/g in fed-batch fermentation at pH 9.0, reducing the risk of contamination during fermentation; and (iii) in thermal fermentation (≥50 °C), *Bacillus* sp. can produce lactic acid. These characteristics should give *Bacillus* spp. several advantages over other bacteria. Firstly, the cost of cooling water after medium sterilization would be reduced. Secondly, *Bacillus* spp. allowed open fermentation on non-sterilized media at temperatures higher than 40 °C (Zhao et al. 2010). Other than that, *Bacillus* strains can also grow and ferment lignocellulosic biomass hexoses and pentoses to lactic acid. *B. coagulans* 36D1 has been shown to use pentose sugar via the pentose–phosphate pathway, which maximizes lactic acid yield up to 1.0 g/g-consumed sugar (Patel et al. 2006).

It takes a long time and a lot of effort to find exceptional lactic acid manufacturers from natural sources. As a result, recent research has used engineering methods to improve fermentation efficiency by employing *E. coli* strains, which can quickly metabolize hexose and pentose sugars, have low dietary requirements, and are easy to genetically alter. To accommodate reducing equivalents generated during glycolysis, wild-type *E. coli* typically produces a mixture of ethanol and several organic acids (lactic acid, acetic acid, succinic acid, and formic acid) (Zhou et al. 2003).

E. coli strains have been engineered in the field of metabolic engineering to increase lactic acid generation (Chang et al. 1994; Zhou et al. 2003). Several studies have used modified E. coli strains to produce lactic acid from glucose (Chang et al. 1994; Dien et al. 2001; Zhou et al. 2005; Zhu et al. 2007), xylose (Dien et al. 2001), sucrose (Wang et al. 2012; Zhou et al. 2005), and glycerol (Mazumdar et al. 2010). However, the productivity (≤1.04 g/L/h), final concentration (≤62.5 g/L), and lactic acid tolerance of engineered *E. coli* strains were significantly lower than that of many LAB and *Bacillus* spp. (Chang et al. 1994; Portnoy et al. 2008; Zhou et al. 2003). As a result, the primary issue to be addressed is achieving high-efficiency lactic acid production using low-cost resources.

C. glutamicum is a Gram-positive, fast-growing, aerobic, nonsporulating, nonmotile, saprophytic bacterium that has been shown to excrete a small amount of several organic acids under oxygen-limited conditions (Yukawa et al. 2007). Several engineered *C. glutamicum* strains have been reported to produce combined organic acids (lactic acid, succinic acid, and acetic acid) from various sugars such as L-arabinose (Kawaguchi et al. 2008; Sasaki et al. 2009), D-glucose (Kawaguchi et al. 2006 2008; Sasaki et al. 2008), D-xylose (Sasaki et al. 2008) and D-cellobiose (Sasaki et al. 2008) when oxygen is depleted. Jia and team reported that by knocking of the L-LDH gene and heterologous expression of the D-LDH encoding gene from *Lactobacillus bulgaricus*, engineered *C. glutamicum* able to produce 17.9 g/L of D-lactic acid (with optical purity > 99.9%) after 16 h of fermentation, which was 32.3% higher than the lactic acid production of the parental strain (Jia et al. 2011). The use of *Corynebacteria* under oxygen-deprived conditions at a high cell density (HCD) is advantageous because energy is primarily directed toward lactic acid production rather than cell growth (Inui et al. 2004). Ica et al. (2005) found a direct relationship between cell concentration and acid production rates, even at high cell densities, and obtained L-lactic acid productivities of 42.9 g/L/h in the mineral medium for more than 360 h at a cell concentration of 60 g dry cells/L. As a result, *C. glutamicum* is an appropriate bacterium capable of high lactic acid production without the need for complex nutritional media.

One of the most important factors in the economic production of lactic acid is the cost of raw materials. Pure sugars or edible crops have traditionally been used as a substrate for lactic acid production, which is advantageous in terms of obtaining a pure lactic acid product while lowering pretreatment and recovery costs. Since process scale-up cannot reduce substrate costs, extensive research is currently being conducted to identify novel substrate for lactic acid production. In that sense, agricultural byproducts, food industry byproducts, and natural unutilized biomasses such as starchy biomass, lignocellulosic biomass, whey, yoghurt, glycerol, and algal biomass have all been explored as potential alternative substrates and renewable resources for lactic acid production. Table 1 summarizes some of the alternative substrates used for lactic acid production.

Other than substrate, fermentation process also affects lactic acid production. The type and nature of substrate, microbial growth, and viscosity of fermentation broth influenced fermentation process selection. Batch, fed-batch, repeated, and continuous fermentations are the examples of fermentation process of lactic acid. Each of the process has its own advantages and disadvantages and are summarized in Table 2 below.

Batch fermentation is the most basic and widely used fermentation process, in which no carbon substrates or other components are added during fermentation, except for neutralizing agents for pH control. In comparison to other fermentation methods, this closed system has advantages in terms of lowering the risk of contamination and obtaining high lactic acid concentrations (Hofvendahl & Hahn–Hägerdal 2000). Batch fermentation, on the other hand, suffers from low cell concentrations due to limited nutrient levels and low productivity because of substrate and/or product inhibition. Kinetic studies on lactic acid production revealed that increasing the initial

Table 1 Pros and cons of lactic acid fermentation modes

Fermentation mode	Advantages	Disadvantages
Batch	Simple operation Reduced risk of contamination High product concentration	Substrate and/or end-product inhibition Low productivity
Fed-batch	High product concentration Overcome substrate inhibition problem	Difficult to achieve optimal design End-product inhibition
Repeated	Time and labor-saving process Short main culture High growth rates Omit seed preparation time	Require special devices or connection line used for cell concentration
Continuous	Less shutdown process frequency Controlled growth rate High productivity	Incomplete utilization of carbon source

Table 2 Lactic acid yield from alternative substrate

Alternative substrate	Microbes used	Lactic acid concentration (g/L)	Lactic acid yield (g/g)	References
Alfalfa fibers	Lactobacillus delbreuckii	35.4	0.35	Sreenath et al. (2001)
	Lactobacillus plantarum	46.4	0.45	
Apple pomace	Lactobacillus rhamnosus ATCC 9595 (CECT 288)	32.5	0.88	Gullón et al. (2008)
Cassava bagasse	Lactobacillus delbreuckii NCIM 2025	81.9	0.94	John et al. (2007)
	Lactobacillus rhamnosus and Bacillus coagulans	112.5	0.88	Chen et al. (2020)
Kitchen waste	Lactic acid bacteria and Clostridium sp.	64.0	0.62	Zhang et al. (2008)
Mango peel	Indigenous microorganism	17.4	–	Jawad et al. (2013)
Mussel processing waste	Lactobacillus plantarum A6	8.4	0.98	Pintado et al. (1999)
Paper sludge	Lactobacillus rhamnosus ATCC 7469	73.0	0.97	Santos et al. (2008)
Sugarcane bagasse	Lactococcus lactis IO-1	10.9	0.36	Laopaiboon et al. (2010)
Waste sugarcane bagasse	Lactobacillus delbreuckii mutant Uc-3	67.0	0.83	Adsul et al. (2007)
Wheat straw	Lactobacillus brevis CHCC 2097 Lactobacillus pentosus CHCC 2355	7.1	0.95	Garde et al. (2002)

glucose concentration by as much as 200 g/L, increases the final lactic acid concentration (Gon et al. 1992; Kadam et al. 2006). The highest lactic acid concentration—up to 192 g/L of lactic acid from 200 g/L of glucose obtained from batch fermentation was reported by Moon et al. (2012) where in their fermentation process, *Lactobacillus paracasei* subsp. *paracasei* CB2121 was used as the fermentation microbes.

Substrate concentrations above the critical level cause substrate inhibition on strains, such as cell lysis and a long lag phase, resulting in a decrease in fermentation rate and sugar utilization (Ding & Tan 2006). To minimize substrate inhibition, fed-batch fermentation is thought to be a better fermentation system for maintaining substrate concentration at a low level compared to batch and continuous fermentation by means of feeding nutrients to the fermentation broth (Abdel-Rahman et al. 2011). In fed-batch fermentation, the substrate is fed to the fermenter continuously or sequentially without the removal of the fermentation broth (Ding & Tan 2006). This method is especially useful when high substrate concentrations have an impact on cell growth and productivity (Lee et al. 1999). It does not, however, resolve severe product inhibition caused by an accumulation of excessive lactic acid product. Although the fed-batch process is an industrially preferred fermentation method because it allows for higher lactic acid concentrations up to 226 g/L as reported by Wang et al. (2011), the productivity is relatively low. Thus, several factors, including the times and terms of feeding the substrate, the substrate concentration to be maintained in the fermentation broth, and how to feed the substrate fermentation (the feeding methods), should be taken into consideration to maximize product concentration in fed-batch fermentation.

Repeated fermentation with batch or fed-batch fermentation involves repeated cycles by inoculating a part or all of the cells from a previous run into the next run (Zhao et al. 2010). Cell recycling methods for bacteria included centrifugation, (Nolasco-hipolito et al. 2012; Zhao et al. 2010), the use of hollow fiber modules (Oh et al. 2003; Wee et al. 2006) and using part of the culture (Zhao et al. 2010). A maximum L-lactic acid production of 107 g/L with an optical purity of 99.8% using a thermophilic *Bacillus* sp. strain 2–6 in open (non-sterilized) repeated batch fermentation was reported by Zhao and team (Zhao et al. 2010). In another study reported by Kim et al. (2006), *Lactobacillus* sp. RKY2 improved lactic acid productivity (6.34 g/L/h) in repeated batch fermentation from cheese whey, which was 6.2 times higher than that obtained in batch fermentation. In comparison to batch or fed-batch fermentation, repeated fermentation has several advantages, including increased yield, time and labor savings for cleaning and sterilizing the fermenter, elimination of seed preparation time, high cell concentration and lactic acid productivity, and short fermentation time due to the high initial inoculation volume (Naritomi et al. 2002).

Lactic acid production is tightly linked to cell growth, allowing cells to get the energy they require for growth via lactic acid-producing pathways. By diluting the product in the fermentation broth with fresh medium, continuous lactic acid fermentation avoids end-product inhibition that occurs in batch/fed-batch fermentation (Amrane & Prigent 1996; Wee & Ryu 2009). Chemostat fermentation is a type of continuous fermentation system in which fresh medium is fed into the fermenter and fermentation broth is withdrawn at the same rate to provide constant control of a component's concentration in the fermentation broth. During chemostat fermentation, the concentrations of cells, products, and substrates in the fermentation broth can be kept stable at constant levels. The specific growth rate can be adjusted by the dilution rate because the specific growth rate would equal the dilution rate under steady state conditions in chemostat fermentation (Bustos et al. 2007). As a result, the dilution rate is an important component for maximizing productivity and should be examined further.

In conventional continuous fermentation without HCD, an improved lactic acid productivity of 1.56 g/L/h from sago starch with the amylolytic LAB strain *Enteroccocus faecium* No. 78 (Shibata et al. 2007) and of 4.53 g/L/h from cassava starch by *Lactobacillus plantarum* SW14 (Bomrungnok et al. 2012) was reported. Bustos and team (2007) reported a lactic acid productivity of 3.1 g/L/h from trimming vine shoots hydrolysate at a dilution rate of 0.2 h − 1 by *Lactobacillus pentosus* CECT-4023 T. D-lactic acid productivities of 2.07–3.55 g/L/h in conventional continuous culture was reported, which was higher than that obtained in the batch culture (1.67 g/L/h) from glucose by *Lactobacillus delbrueckii* subsp. *lactis* QU 41 (Tashiro et al. 2011). Furthermore, continuous fermentation is shut down less frequently than batch fermentation (where the reactor must be emptied, cleaned, sterilized, and refilled), and productivity decreases less during lag phases (Gassem et al. 1997).

Cost effective lactic acid production can be achieved by choosing fermentation substrate from either plant wastes or domestic wastes. Coupled with fermentation process that gives high productivity as well as high product concentration, lactic acid fermentation can be well improved. Since plant wastes are usually originated from rural areas, the usage of it as a feedstock of major fermentation is a part of the circular economy improvement and subsequently contributes to the positive increment of the local economy by creating more job vacancies. The method of reusing wastes to produce another product (in this case, lactic acid) can also effectively cut waste treatment costs and minimizing pollution, thus improving pollution contributed by the fermentation industry.

2.2 *Polyhydroxyalkanoates (PHA)*

Polyhydroxyalkanoates (PHAs) are natural polyesters produced by bacteria for energy storage (Sudesh et al. 2000). Due to similar properties with conventional plastic and biocompatibility with mammalian tissue, this biopolymer group is widely studied as an alternative to conventional plastics (Aslan et al. 2016; Saratale et al. 2021). It is one of the few biopolymers thoroughly studied to replace petroleum-based polymers due to their superiority in terms of biodegradability and physico-chemical similarities with plastics. The major obstacle in the commercialization of PHAs is the production cost (Shogren et al. 2019). A notable decrease in the PHA production cost was observed in the recent years due to enhanced production efficiency. Table 3 summarizes the PHA production strategies. A significant cost was cut-down by utilizing agro-industrial wastes as inexpensive/free substrate sources for the biopolymer production (Kaur et al. 2017), as typically, about half of the production expense is incurred by the utilization of pure carbon source feedstock (Aslan et al. 2016). With the ongoing advancements in research and development, the sustainable and eco-friendly production of PHA with reduced cost is a promising trend with global commercial interest.

Among the existing methods of PHA production, microbial fermentation is the prime and prominent method studied for PHA synthesis (Saratale et al. 2021). The recent advancements in microbial fermentation to produce PHA were highlighted in the work by Koller (2018). It is notable that the PHA type and production rate varies

Table 3 Strategies for efficient PHA biosynthesis

Reference(s)	Production Strategy	Notable outcome	Limitation
Chen and Jiang (2018); Zou et al. (2017)	Genetic engineering	Utilizing extremophile microbes and developing extremophile PHA synthases for enhanced PHA biosynthesis	Lack of experimental designs and optimization of PHA production parameters
Oliveira et al. (2019); Sabapathy et al. (2020); Tu et al. (2019)	Mixed culture method	With no intensive sterilization requirements, efficient low-cost PHA productions were achieved	Maximizing PHA content for downstream processing requires further research
Blunt et al. (2019)	High cell density (HCD) cultivation	O_2 limitation promoted by HCD cultures showed an efficient biosynthesis of PHA by utilizing fatty acids	A lower carbon recovery was noted from feedstock to PHA yield
Mamat et al. (2020); Penloglou et al. (2012); Tu et al. (2019)	Mathematical modeling	Employing mathematical modeling tools and statistical programs, efficient utilization of resources for faster maximum yield, its prediction and monitoring can be facilitated	Thorough validation experiments are required to correlate the model output with real-time results
Gahlawat (2019); Heng et al. (2017); Surendran et al. (2020)	Renewable substrates	Significant cost reduction due to renewable substrates is a major progress in the economic viability of the PHA commercial application	A systematic study on specific plant waste and the most suitable microbe for PHA biosynthesis must be carried out
Hawkins et al. (2021)	Biofloc technology (BFT)	BFT on aquaculture provides a new insight on PHA cultivation for large-scale commercial production	Few studies are available on the process optimization of the BFT for PHA production

based on the employed substrate and microbe. The fermentation process to produce PHA is led by three core pathways that require fatty acids and sugars selectively (Saratale et al. 2021). The metabolic pathways and their significance in PHA synthesis have been exhaustively discussed by Choi et al. (2020). The need for sugars and fatty acids in the pathways paves way for exploring plant-based waste utilization that are rich in those nutrients, as the feedstock cost holds the most weight in PHA production (Aslan et al. 2016; Chen & Jiang 2018). Careful inspection of PHA production strategies is the key step in devising the ultimate optimized pathway for its sustainable commercialization. The prominent PHA production strategies and insights are summarized in Table 3.

As the PHA production is biosynthesized by microbial fermentation, genetic engineering plays an important role in maximizing the biopolymer yield. Establishing genetically modified proteins, especially, enzymes for PHA synthesis, and exploring extremophile bacteria that can highly function in adverse conditions are of major interest in propelling the sustainability of PHA production (Chen & Jiang 2018; Zou et al. 2017). Besides, mixed culture method technique for the microbial fermentation PHA production is significantly enhanced. As the chosen diverse bacterial cultures compete for the specified limited feedstock, at the presented feast and famine conditions, an enhanced PHA production is a notable outcome, especially without the need of intensive sterilization (Oliveira et al. 2019; Sabapathy et al. 2020; Tu et al. 2019). Hence, the costs and complexities of the process are reduced. Moreover, limiting oxygen in a HDC with increased utilization of fatty acids, resulted in an efficient PHA biosynthesis (Blunt et al. 2019). Also, a new approach employing BFT on aquaculture, indicated a promising insight on large-scale PHA production, as an improved solution for cost effectiveness and commercialization (Hawkins et al. 2021).

A common obstacle in most of these processes is the lack of mathematical modeling. Employing powerful statistical and mathematical modeling software for the optimization of substrates, culture performance, and other critical operational parameters and understanding the interaction with the response variables, can greatly reduce the amount of experiments needed with minimum resources (Mamat et al. 2020; Penloglou et al. 2012; Tu et al. 2019). Although employing genetic engineering, HCD and mixed culture method approach, mathematical modeling, and BFT, the most notable source of cost is the carbon feedstock. Nearly half of the PHA production costs are incurred by the feedstock required for PHA biosynthesis (Aslan et al. 2016). As carbon rich plants can be used for providing renewable substrates for the PHA biosynthesis, a rising competition with food sources makes the process unsustainable. However, huge amounts of plant wastes with unused carbon, are a potential carbon sources to be utilized for PHA biosynthesis tackling not only the high expenses from carbon feedstock, but also minimizing pollution by converting waste to wealth. Hence, a circular economy can be established. Aslan et al. (2016) reviewed the potential of various agricultural waste materials for PHA production. Besides, a tremendous progress is achieved in cutting down production costs with renewable substrates (Gahlawat 2019; Heng et al. 2017; Surendran et al. 2020).

To employ plant waste as feedstock, necessary pretreatment is required to harness fermentable sugars. The pretreatment steps every biomass waste feedstock goes through, are generally drying for moisture removal, pulverization, and hydrolysis with the support of acid, alkali, or enzyme. The physical, chemical, or biochemical pre-processing, followed by the type of hydrolysis, depends on the type of plant waste. The resulting hydrosylates are the carbon source units for the bacterial fermentation for PHA synthesis.

Popular plant wastes that had been successfully employed for PHA biosynthesis are cassava waste (Salgaonkar et al. 2019), sugarcane bagasse (Getachew & Woldesenbet 2016), sugarcane molasse (Dalsasso et al. 2019), rice husk (Heng et al. 2017) and oil palm biomass (Yustinah et al. 2019) also known as empty fruit bunch (EFB). Depending on the structural differences, the pretreatment of the plant wastes vary, along with the type of bacteria employed for PHA biosynthesis. Table 4 summarizes the notable works on plant waste utilization for PHA production in terms of the specific pretreatment, microbe used, and notable outcome of the process. Figure 1 shows the general overview of the processes involved from plant waste utilization to PHA biosynthesis.

As a huge amount of plant wastes is produced annually worldwide, there arises an opportunity for their sustainable conversion to useful biopolymers (PHA). The effective utilization of plant waste to significantly diminish PHA production costs, is a promising step toward a circular economy, as the waste is not only reduced but also converted to wealth, by commercialization of the bioproduct. Moreover,

Table 4 Summary of notable works on PHA synthesis from plant waste

Reference(s)	Plant waste	Pretreatment method	Microbe used for fermentation	Outcome
Salgaonkar et al. (2019)	Cassava waste	Drying, pulverization followed by acid hydrolysis	Halogeometricum borinquense strain E3	The culture E3 was able to achieve a PHA concentration of 1.52 $g L^{-1}$
Getachew and Woldesenbet (2016)	Sugarcane bagasse	Drying, shredding, and zinc chloride hydrolysis	Bacillus spp	56% PHA (in butyrate form) was achieved
Dalsasso et al. (2019)	Sugarcane molasse	Enzymatic hydrolysis	*Cupriavidus necator*	40 wt% of PHA was obtained at shake flask scale
Heng et al. (2017)	Rice husk	Alkali treatment followed by enzymatic hydrolysis	Burkholderia cepacia USM (JCM 15,050)	50% PHA content was achieved
Yustinah et al. (2019)	Oil-palm biomass	Acid hydrolysis	*Bacillus cereus* suaeda B-001	55.4% PHA accumulation was achieved

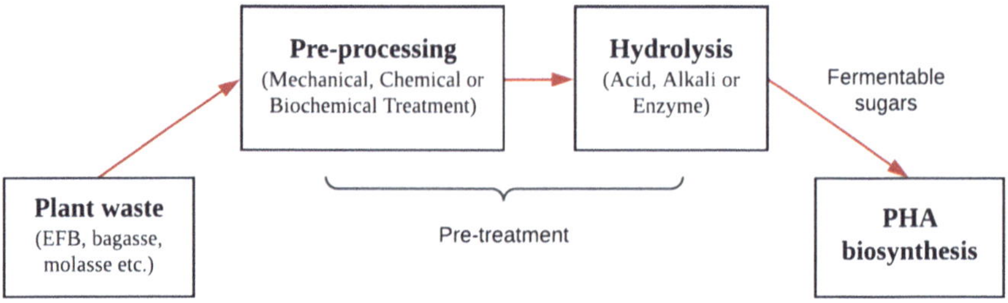

Fig. 1 Overall flowchart of PHA production from plant waste (Sirohi et al. 2020)

the ability to replace conventional non-biodegradable plastics, makes PHA a valuable green biopolymer. Besides, encouraging rural investments for renewable PHA production to minimize transport costs, along with increased job opportunities for the agricultural human resource sector greatly heads toward a promising circular economy (Golden et al. 2018). Moreover, PHA biopolymer-based plastics depicted a significant reduction in greenhouse gas emissions compared to the conventional petroleum-based plastics (Spierling et al. 2018). Furthermore, PHA nanocomposites greatly contribute to the medical field in terms of stem cell engineering, drug carrier, scaffold materials, and more, beside safe food packaging material (Saratale et al. 2021).

PHA is a biopolymer with great commercial interest along with sustainable outcomes in environmental and medical fields. The biggest barrier to market entry of the PHA biopolymer-based products is the production cost. A huge portion of the PHA biosynthesis expense comes from the carbon source feedstock. Therefore, the potential for plant waste utilization for PHA production substrates along with enhanced PHA production strategies, show a promising potential toward the development of a circular economy, not only supporting the environment by reducing pollution and greenhouse gas emissions, but also opening job opportunities along with rural investments.

2.3 Carboxymethyl Cellulose (CMC)

Cellulose is a large type of natural polymers which their production exceeds 10^{12} tons per year (Kim et al. 2021). It is a homo-biopolymers with a high molecular weight unbranched chain with repeating D-glucose units, linked together by ß-1,4 glycosidic bonds. It has been produced by replacing the hydroxyl groups of cellulose with methyl group to increase the utility of cellulose. Carboxymethyl cellulose (CMC) is one of the commonly high demanded cellulose derivatives due to vast advantages such as biodegradability, biocompatibility, and most abundance especially can easily obtained from agricultural waste. According to Oun et al., CMC has a good film-forming properties, good optical transparency, and chemically stable (Oun & Rhim

2017). Owing to these valuable biopolymers can be found abundantly from agricultural wastes, cellulose-derivatives are the potential products to be exploited as well as turning the agricultural wastes to wealth.

Scientists classify the synthesis of CMC into two (2) ways; conventional way which utilizes the plant based as the CMC precursor and non-conventional way are from waste materials from household and office products. Therefore, this subchapter focuses on the conventional ways to highlight the synthesis of CMC from agricultural wastes. Table 5 summarizes the recent studies on the synthesis of CMC and the properties of CMC produced from various sources of agricultural wastes. The physical properties of CMC and the parameter of fabrication also influenced by the method of extraction and the precursors which are the sources of agricultural wastes (Maraveas 2020). The substitution of carboxymethyl group with hydroxyl groups usually happened at C-2 glucose position (Charpentier et al. 1997).

Cellulosic material shows a great promising biomaterial owing to their biocompatibility, biodegradability, low toxicity, and the most important cost-effective. CMC is one of the important products of cellulose ethers (Chen 2015). As aforementioned, CMC nowadays obtain a great attention among researchers and highly demand for the environmental-friendly company that aiming to meet the sustainable development goals (SDGs). Due to that, a lot of potential applications of CMC have been explored and summarized in Fig. 2. The major applications are in the medical and pharmaceutical industries such as in tissue engineering, wound dressing, fabrication of 3D-scaffolds biocompatible implants, artificial organs, drug delivery, and stabilization purposes (Rahman et al. 2021). In drug delivery, selection of good carrier is important to help the drug to be delivered at the targeted location, reduce fast absorption that led to toxicity as well as sustain the release of drugs. Rahman et al. (2021) mentioned in his review that CMC was used to encapsulate the docetaxel (DTX) resulting the best carrier for this anti-cancer chemotherapy drugs. Besides, the application of CMC also for water treatment (Saber-Samandari et al. 2016). Not to mention, CMC was also reported for applications in manufacturing.

2.4 Gum Arabic (GA)

Natural plant polymers are becoming more prevalent in the industrial realm and scientific research centers recently. Furthermore, research shows that health consciousness is steadily increasing in modern societies nowadays, increasing the demand for these valuable natural polymers and amplifying their various applications. Gum Arabic (GA) is a plant exudate harvested from the *Acacia senegal* and *Acacia seyal* species which belong to *Fabaceae (Leguminosae)* family. GA is considered one of the most ancient and well-known among the natural gums and it thrives and harvested in sub-Sahara African countries (Awad et al. 2018).

Moreover, its properties of "Generally Recognized as Safe (GRAS)" status, edibility, and organoleptic acceptance "taste & odorless", adding to that its excellent solubility in water, attracted major global industries and marketing platforms in the

Table 5 Sources of cellulose extraction and the method of producing CMC

References	Agricultural wastes	Pretreatment of raw materials	Method of production of CMC	Outcomes
Haqiqi et al. (2021)	Oil-palm empty fruit bunch (EFB)	Alkaline extraction, Bleaching,	Modified the cellulose pulp with NaOH before the addition of sodium monochloroacetate	CMC was obtained readily 81.32% water-soluble
Kim et al. (2021)	Hardwood bleached pulp	Pretreatment with endoglucanase	Prepared by solution casting method	CMC film obtained with 75.2 ± 4.2 μm thickness was thickest compared to cellulose nanofibers (CNF) in this study
Alizadeh et al. (2017)	Sugarcane bagasse	Pretreatment with sodium hypochlorite, chlorine gas, and KCL to produce cellulose	Same method as Rachtanapun et al., The cellulose bagasse powder was mixed with a variety of concentration of NaOH and IPA as a solvent. After stirring, sodium monochloroacetate was added	Elongation break and the degree of substitution (DS) of CMC increase with increasing the concentration of NaOH but not more than 40% concentration of NaOH
Haleem et al. (2014)	Cotton gin	Pretreatment of cotton gin with HCl and H_2SO_4 to obtain cellulose	The CMC was produced by etherification using monochloroacetic acid and NaOH	The optimum production of CMC from the cotton gin was obtained by 10 g cotton gin in 100 mL NaOH that providing highest viscosity and 0.874 degree of substitutions
Rachtanapun et al. (2012)	Durian rind	Pretreatment with durian rind powder in 10 g/100 mL NaOH and bleached with hydrogen peroxide to produce cellulose	The conversion of cellulose into CMC was done by carboxymethylation using sodium monochloroacetate and NaOH	The optimum condition for CMC fabrication was found in 30 g of dried durian rind in 100 mL NaOH that obtained 0.87 of degree of substitution

(continued)

Table 5 (continued)

References	Agricultural wastes	Pretreatment of raw materials	Method of production of CMC	Outcomes
Rachtanapun et al. (2010)	Papaya peels	Treatment with 0.5 M NaOH to produce cellulose	Addition of chloroacetic acid from cellulose powder and undergoes precipitation, neutralization, washing, and filtration until obtain final product, CMC	Density of ceramic slips decreased with increasing of CMC concentrations. As compared to commercial CMC, extracted CMC from papaya gave higher bending strength
Adinugraha et al. (2005)	Banana pseudo stem	Pretreatment was done by bleaching with 5% of sodium hypochlorite for 3 h at 30 °C	The CMC was obtained by alkalinizing of cellulose with 2-mL of various concentration of NaOH in 100 mL isopropanol as the solvent and then sodium monochloroacetate was added	15% NaOH and etherification using 1.2 g w/w NaMCa obtained the highest degree of substitution (0.75), 4033 cps viscosity, 98.23% purity, and 38.33% crystallinity

food and pharmaceutical industries. GA is still used widely due to its texture, film-forming, emulsifying, and stabilizing properties despite the periodical attempts to replace it (Adeyanju & Akomolafe 2013).

Some of the prevalent common uses of this polymer are as an emulsifier, stabilizing, thickening, and binding agent in industrial processes, such as food, pharmaceutical, cosmetics, adhesives, paints, inks, lithography, and textile industries (Montenegro et al. 2012) adding to that cardio-vascular, renal, gut, and dental- protective, satiety-inducing, antimicrobial, anti-inflammatory activities reported in current literature (Patel & Goyal 2015).

Moreover, GA is becoming a promising candidate as an encapsulating agent or coating material in general, and particularly for essential oils "EOs" (Patel & Goyal 2015). GA polymer consists of about 90 to 99% of Arabinogalactan, whereas Glycoprotein represents the remaining 1 to 10% (Gulão et al. 2014). Nevertheless, the GA can establish an excellent and stable emulsion with EOs (Aphibanthammakit et al. 2020), making it a good potential coating polymer for essential oils in the encapsulation process. GA is composed of arabinogalactan (AG), arabinogalactan protein (AGP), and glycoprotein (GP). However, the AGP is the one that mostly contributing emulsifying property of the polymer (Atgié et al. 2019).

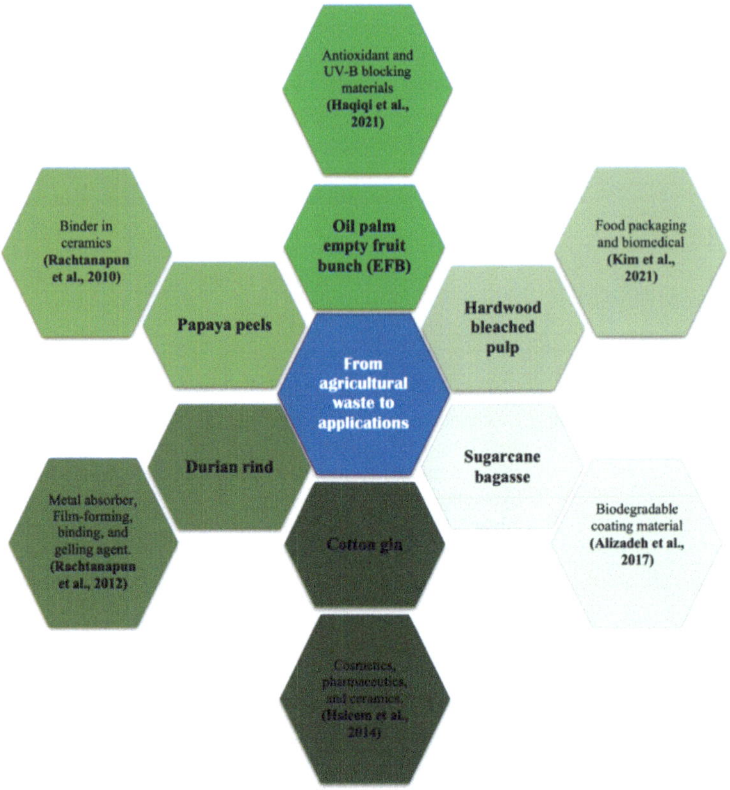

Fig. 2 Recent studies on CMC-derived from agricultural waste and their potential applications

2.4.1 *Acacia Seyal* (Talha) and *Acacia Senegal* (Hashab)

These species "*A. Seyal and A. Senegal*" are naturally grown in the gum belt region of Sub-Saharan African countries, such as Sudan, Chad, and Nigeria which are the major gum producers internationally (Odee et al. 2012). However, *A. Seyal* gum is considered less quality than *A. Senegal* in terms of gum quantity (Awad et al. 2018). Figure 3 shows the Gum Arabic available in the market and the chemical structure is shown in Fig. 4.

2.4.2 Characterization of GA

The trees of these two species (*A. Senegal and A. Seyal*) are easily distinguishable; however, their gum product (GA) is quite similar and usually characterized by two main properties; their molar mass difference, which occur between (2.8×10^5 and 2.5×10^6 g. mole 1) and their protein content 3% and 1% respectively. Hence, *A. Senegal*

Fig. 3 Gum Arabic

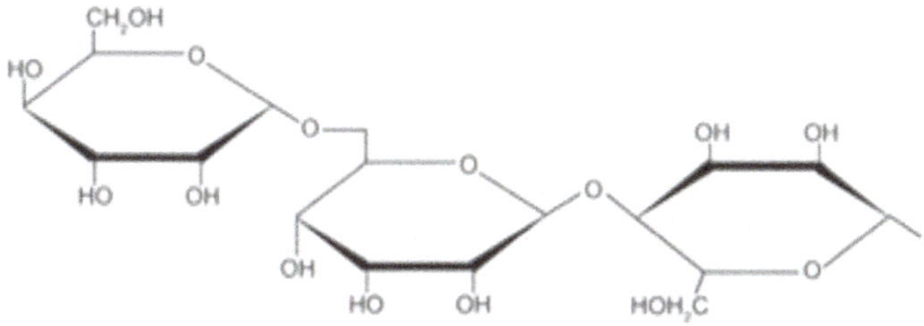

Fig. 4 Gum Arabic structure

is more efficient for emulsion preparation as far as emulsion stability is concerned. However, the AGP concentration and its structural diversity between the two species are considered the major contributor to the quality differences (Aphibanthammakit et al. 2020).

2.4.3 GA Applications

The GA biopolymer is found in many different industrial applications, including. Food additives, shelf-life enhancers, microencapsulates, (Maqbool et al. 2011; Mosquera et al. 2012; Su et al. 2008) as an emulsifier it was found to provide stability over a wide pH range, compared to other emulsifiers (Su et al. 2008). GA could be used as an alternative to paraffin wax for fruits and vegetable coating; Maqbool et al. suggested that 10% of GA and 1% chitosan can be used to make an edible coating, commercially suitable for extending the storage life of bananas up to 33 days (about 1 month).

3 Conclusion

With advancements in the chemical and petrochemical production industries, remarkable economic boost has been harvested to meet global demands. However, there is a rising concern regarding the resulting environmental pollution and scarcity of resources, that requires strict consideration for the long-term wellbeing of the ecosystem. Everything in nature inspires a cyclic nature of events starting from the growth of a plant from soil to fruit harvesting and the wastes returned to the soil again, enriched with decomposed nutrients for further growth cycles of vegetation. The sustainability of nature maintained with its cyclic renewable patterns is the inspiration for adopting and implementing the principles in all aspects of life, particularly the circular economy. Hence, utilizing carbon rich plant wastes for useful eco-friendly biomaterials paves the way for a sustainable circular economy incorporating the idea of turning waste to wealth.

References

Abdel-Rahman, M. A., Tashiro, Y., & Sonomoto, K. (2011). Lactic acid production from lignocellulose-derived sugars using lactic acid bacteria: Overview and limits. *Journal of Biotechnology, 156*(4), 286–301. https://doi.org/10.1016/j.jbiotec.2011.06.017

Abdel-Rahman, M. A., Tashiro, Y., & Sonomoto, K. (2013). Recent advances in lactic acid production by microbial fermentation processes. *Biotechnology Advances, 31*(6), 877–902. https://doi.org/10.1016/j.biotechadv.2013.04.002

Adeyanju, O., & Akomolafe, S. F. (2013). Extraction and physiochemical characterization of a new polysaccharide obtained from Daniella oliveri (Rolfe) Hutch and Dalz (LEGUMINOSAE). In: *Proceeding of the 36th annual international conference of chemical society of Nigeria, Minna, Nigeria* (Vol. 2, pp. 513–520).

Adinugraha, M. P., Marseno, D. W., & Haryadi. (2005). Synthesis and characterization of sodium carboxymethylcellulose from cavendish banana pseudo stem (*Musa cavendishii* LAMBERT). *Carbohydrate Polymers, 62*(2), 164–169. https://doi.org/10.1016/j.carbpol.2005.07.019

Adsul, M. G., Varma, J., & Gokhale, D. V. (2007). Lactic acid production from waste sugarcane bagasse derived cellulose. *Green Chemistry* 58–62. https://doi.org/10.1039/b605839f

Ahmad, A., Banat, F., & Taher, H. (2020). A review on the lactic acid fermentation from low-cost renewable materials: Recent developments and challenges. *Environmental Technology and Innovation, 20*, 101138. https://doi.org/10.1016/j.eti.2020.101138

Alizadeh Asl, S., Mousavi, M., & Labbafi, M. (2017). Synthesis and characterization of carboxymethyl cellulose from sugarcane bagasse. *Journal of Food Processing & Technology, 08*(08). https://doi.org/10.4172/2157-7110.1000687

Amrane, A., & Prigent, Y. (1996). *Biotechnol Lactic Acid, 1656*(95).

Aphibanthammakit, C., Barbar, R., Nigen, M., Sanchez, C., & Chalier, P. (2020). Emulsifying properties *of Acacia senegal* gum: Impact of high molar mass protein-rich AGPs. *Food Chemistry: X, 6*, 100090. https://doi.org/10.1016/J.FOCHX.2020.100090

Aslan, A. K. H. N., Ali, M. D. M., Morad, N. A., & Tamunaidu, P. (2016). Polyhydroxyalkanoates production from waste biomass. *IOP Conference Series: Earth and Environmental Science, 36*(1). https://doi.org/10.1088/1755-1315/36/1/012040

Atgié, M., Garrigues, J. C., Chennevière, A., Masbernat, O., & Roger, K. (2019). Gum Arabic in solution: Composition and multi-scale structures. *Food Hydrocolloids, 91*, 319–330. https://doi.org/10.1016/J.FOODHYD.2019.01.033

Awad, S. S., Rabah, A. A., Ali, H. I., & Mahmoud, T. E. (2018). Acacia seyal Gums in Sudan: Ecology and economic contribution. *Gum Arabic: Structure, Properties, Application and Economics*, 3–11. https://doi.org/10.1016/B978-0-12-812002-6.00001-4

Benthin, S., & Villadsen, J. (1995). Production of optically pure d-lactate by *Lactobacillus bulgaricus* and purification by crystallisation and liquid/liquid extraction. *Applied Microbiology and Biotechnology, 42*(6), 826–829. https://doi.org/10.1007/BF00191176

Blunt, W., Dartiailh, C., Sparling, R., Gapes, D. J., Levin, D. B., & Cicek, N. (2019). Development of high cell density cultivation strategies for improved medium chain length polyhydroxyalkanoate productivity using *Pseudomonas putida* LS46. *Bioengineering, 6*(4). https://doi.org/10.3390/bioengineering6040089

Bomrungnok, W., Sonomoto, K., Pinitglang, S., & Wongwicharn, A. (2012). Single step lactic acid production from cassava starch by *Laactobacillus plantarum* SW14 in conventional continuous and continuous with high cell density. *APCBEE Procedia, 2*, 97–103. https://doi.org/10.1016/j.apcbee.2012.06.018

Budhavaram, N. K., & Fan, Z. (2009). Production of lactic acid from paper sludge using acid-tolerant, thermophilic *Bacillus coagulan* strains. *Bioresource Technology, 100*(23), 5966–5972. https://doi.org/10.1016/j.biortech.2009.01.080

Bustos, G., Moldes, A. B., & Cruz, J. M. (2007). Revalorization of hemicellulosic trimming vine shoots hydrolyzates trough continuous production of lactic acid and biosurfactants by *L. pentosus*. *Journal of Food Engineering, 78*, 405–412. https://doi.org/10.1016/j.jfoodeng.2005.10.008

Calabia, B. P., Tokiwa, Y., & Aiba, S. (2011). Fermentative production of l-(+)-lactic acid by an alkaliphilic marine microorganism. *Biotechnology Letters, 33*(7), 1429–1433. https://doi.org/10.1007/s10529-011-0573-0

Chang, H. N., Yoo, I. K., & Kim, B. S. (1994). High density cell culture by membrane-based cell recycle. *Biotechnology Advances, 12*(3), 467–487. https://doi.org/10.1016/0734-9750(94)90020-5

Charpentier, D., Mocanu, G., Carpov, A., Chapelle, S., Merle, L., & Muller, G. (1997). New hydrophobically modified carboxymethylcellulose derivatives. *Carbohydrate Polymers, 33*(2–3), 177–186. https://doi.org/10.1016/S0144-8617(97)00031-3

Chen, G. Q., & Jiang, X. R. (2018). Engineering microorganisms for improving polyhydroxyalkanoate biosynthesis. *Current Opinion in Biotechnology, 53*, 20–25. https://doi.org/10.1016/j.copbio.2017.10.008

Chen, H., Chen, B., Su, Z., Wang, K., Wang, B., Wang, Y., Si, Z., Wu, Y., Cai, D., & Qin, P. (2020). Industrial crops & products efficient lactic acid production from cassava bagasse by mixed culture of *Bacillus coagulans* and *Lactobacillus rhamnosus* using stepwise pH controlled simultaneous saccharification and co-fermentation. *Industrial Crops & Products, 146*(15), 112175. https://doi.org/10.1016/j.indcrop.2020.112175

Chen, H. (2015). Lignocellulose biorefinery product engineering. *Lignocellulose Biorefinery Engineering*, 125–165. https://doi.org/10.1016/B978-0-08-100135-6.00005-3

Choi, S. Y., Rhie, M. N., Kim, H. T., Joo, J. C., Cho, I. J., Son, J., Jo, S. Y., Sohn, Y. J., Baritugo, K. A., Pyo, J., Lee, Y., Lee, S. Y., & Park, S. J. (2020). Metabolic engineering for the synthesis of polyesters: A 100-year journey from polyhydroxyalkanoates to non-natural microbial polyesters. *Metabolic Engineering, 58*, 47–81. https://doi.org/10.1016/j.ymben.2019.05.009

Dalsasso, R. R., Pavan, F. A., Bordignon, S. E., de Aragão, G. M. F., & Poletto, P. (2019). Polyhydroxybutyrate (PHB) production by Cupriavidus necator from sugarcane vinasse and molasses as mixed substrate. *Process Biochemistry, 85*(May), 12–18. https://doi.org/10.1016/j.procbio.2019.07.007

de Oliveira, G. H. D., Niz, M. Y. K., Zaiat, M., & Rodrigues, J. A. D. (2019). Effects of the organic loading rate on polyhydroxyalkanoate production from sugarcane stillage by mixed microbial cultures. *Applied Biochemistry and Biotechnology, 189*(4), 1039–1055. https://doi.org/10.1007/s12010-019-03051-9

Dien, B. S., Nichols, N. N., & Bothast, R. J. (2001). Recombinant *Escherichia coli* engineered for production of. *Journal of Industrial Microbiology and Biotechnology, 27*(July), 259–264.

Ding, S., & Tan, T. (2006). Short communication L-lactic acid production by *Lactobacillus casei* fermentation using different fed-batch feeding strategies. *Process Biochemistry, 41*, 1451–1454. https://doi.org/10.1016/j.procbio.2006.01.014

Ellen MacArthur Foundation. (2017). The new plastics economy: Rethinking the future of plastics & catalysing action. *Ellen MacArthur Foundation, 68*. https://doi.org/10.1103/Physrevb.74.035409

Flores-Albino, B., Arias, L., Gómez, J., Castillo, A., Gimeno, M., & Shirai, K. (2012). Chitin and L(+)-lactic acid production from crab (*Callinectes bellicosus*) wastes by fermentation of Lactobacillus sp. B2 using sugar cane molasses as carbon source. *Bioprocess and Biosystems Engineering, 35*(7), 1193–1200. https://doi.org/10.1007/s00449-012-0706-4

Gahlawat, G. (2019). *Production strategies for commercialization of PHA* (Issue 2009). https://doi.org/10.1007/978-3-030-33897-8_4

Garde, A., Jonsson, G., Schmidt, A. S., & Ahring, B. K. (2002). Lactic acid production from wheat straw hemicellulose hydrolysate by *Lactobacillus pentosus* and *Lactobacillus brevis*. *Biosource Technology, 81*, 217–223.

Gassem, M. A., Sims, K. A., & Frank, J. F. (1997). Extracellular polysaccharide production by *Lactobacillus delbrueckii* subsp. bulgaricus RR in a continuous fermentor. *LWT-Food Science and Technology, 278*(40), 273–278.

Getachew, A., & Woldesenbet, F. (2016). Production of biodegradable plastic by polyhydroxybutyrate (PHB) accumulating bacteria using low cost agricultural waste material. *BMC Research Notes, 9*(1), 1–9. https://doi.org/10.1186/s13104-016-2321-y

Golden, J., S, Y., Handfield, R., Agsten, B., Khan, L., & True, E. (2018). *Indicators of the U. S. biobased economy*. 74.

Gon, L. M. D., Barreto, M. T. O., Xavier, A. M. B. R., Carrondo, M. J. T., Klein, J., Bioquimica, E., Nova, U., Quimica, T., Experimental, B., Gmbh, G., Landstrasse, I., Republic, F., & Received, G. (1992). Inert supports for lactic acid fermentation—a technological assessment. *Applied Microbiology and Biotechnology*, 305–311.

Gulão, E. da S., de Souza, C. J. F., da Silva, F. A. S., Coimbra, J. S. R., & Garcia-Rojas, E. E. (2014). Complex coacervates obtained from lactoferrin and gum arabic: Formation and characterization. *Food Research International, 65*(PC), 367–374. https://doi.org/10.1016/J.FOODRES.2014.08.024

Gullón, B., Yáñez, R., Alonso, J. L., & Parajó, J. C. (2008). L-Lactic acid production from apple pomace by sequential hydrolysis and fermentation. *Bioresource Technology, 99*(2), 308–319. https://doi.org/10.1016/j.biortech.2006.12.018

Haleem, N., Arshad, M., Shahid, M., & Ashraf, M. (2014). Synthesis of carboxymethyl cellulose from waste of cotton ginning industry. *Carbohydrate Polymers, 113*, 249–255. https://doi.org/10.1016/j.carbpol.2014.07.023

Haqiqi, M. T., Bankeeree, W., Lotrakul, P., Pattananuwat, P., Punnapayak, H., Ramadhan, R., Kobayashi, T., Amirta, R., & Prasongsuk, S. (2021). Antioxidant and UV-blocking properties of a carboxymethyl cellulose-lignin composite film produced from oil palm empty fruit bunch. *ACS Omega, 6*(14), 9653–9666. https://doi.org/10.1021/acsomega.1c00249

Hawkins, S., de Fonseca, I. B. F. C, Lima Da Silva, R., & Quirino, R. L. (2021). Aquaculture waste: Potential synthesis of polyhydroxyalkanoates. *ACS Omega, 6*(4), 2434–2442. https://doi.org/10.1021/acsomega.0c05304

Heng, K. S., Hatti-Kaul, R., Adam, F., Fukui, T., & Sudesh, K. (2017). Conversion of rice husks to polyhydroxyalkanoates (PHA) via a three-step process: Optimized alkaline pretreatment, enzymatic hydrolysis, and biosynthesis by Burkholderia cepacia USM (JCM 15050). *Journal of Chemical Technology and Biotechnology, 92*(1), 100–108. https://doi.org/10.1002/jctb.4993

Hofvendahl, K., & Hahn–Hägerdal, B. (2000). Factors affecting the fermentative lactic acid production from renewable resources 1. *Enzyme and Microbiol Technology, 26*, 87–107.

Ica, B., Ucts, P., Pro, N. D., Engine, C., Okino, S., Inui, M., & Yukawa, H. (2005). Production of organic acids by *Corynebacterium glutamicum* under oxygen deprivation. *Applied Microbiology and Biotechnology*, 475–480. https://doi.org/10.1007/s00253-005-1900-y

Inui, M., Murakami, S., Okino, S., Hideo, K., Vertès, A. A., & Yukawa, H. (2004). Metabolic analysis of corynebacterium glutamicum during lactate and succinate productions under oxygen deprivation. *Journal of Molecular Microbiology and Biotechnology, 0292*, 182–196. https://doi.org/10.1159/000079827

Jawad, A. H., Alkarkhi, A. F. M., & Norulaini, N. A. N. (2013). Production of the lactic acid from mango peel waste—Factorial experiment. *Journal of King Saud University - Science, 25*(1), 39–45. https://doi.org/10.1016/j.jksus.2012.04.001

Jia, X., Liu, P., & Li, S. (2011). D-lactic acid production by a genetically engineered strain *Corynebacterium glutamicum. World Journal of Microbiology and Biotechnology*, 2117–2124. https://doi.org/10.1007/s11274-011-0675-9

John, R. P., Nampoothiri, K. M., & Pandey, A. (2007). Production of L (+) lactic acid from cassava starch hydrolyzate by immobilized *Lactobacillus delbrueckii. Journal of Basic Microbiology*, 25–30. https://doi.org/10.1002/jobm.200610208

Kadam, S. R., Patil, S. S., Bastawde, K. B., Khire, J. M., & Gokhale, D. V. (2006). Strain improvement of Lactobacillus delbrueckii NCIM 2365 for lactic acid production. *Process Biochemistry*, 41, 120–126. https://doi.org/10.1016/j.procbio.2005.06.007

Kaur, L., Khajuria, R., Parihar, L., & Dimpal Singh, G. (2017). Polyhydroxyalkanoates: Biosynthesis to commercial production—a review. *Journal of Microbiology, Biotechnology and Food Sciences*, *6*(4), 1098–1106. https://doi.org/10.15414/jmbfs.2017.6.4.1098-1106

Kawaguchi, H., Verte, A. A., Okino, S., Inui, M., Yukawa, H., Yukawa, H., & Mol, J. (2006). Engineering of a xylose metabolic pathway in *Corynebacterium glutamicum. Applied and Environmental Microbiology*, *72*(5), 3418–3428. https://doi.org/10.1128/AEM.72.5.3418

Kawaguchi, H., Sasaki, M., & Vertès, A. A. (2008). Engineering of an L-arabinose metabolic pathway in *Corynebacterium glutamicum. Applied Microbiology and Biotechnology*, 1053–1062. https://doi.org/10.1007/s00253-007-1244-x

Kim, H. J., Roy, S., & Rhim, J. W. (2021). Effects of various types of cellulose nanofibers on the physical properties of the CNF-based films. *Journal of Environmental Chemical Engineering, 9*(5), 106043. https://doi.org/10.1016/j.jece.2021.106043

Kim, H. O., Wee, Y. J., Kim, J. N., Yun, J. S., & Ryu, H. W. (2006). Production of lactic acid from cheese whey by batch and repeated batch cultures of Lactobacillus sp. *RKY2, 129*(1), 694–704.

Koller, M. (2018). A review on established and emerging fermentation schemes for microbial production of polyhydroxyalkanoate (PHA) biopolyesters. *Fermentation, 4*(2). https://doi.org/10.3390/fermentation4020030

Kyla-Nikkila, K., Hujanen, M., Leisola, M., & Palva, A. (2000). Metabolic engineering of *Lactobacillus helveticus* CNRZ32 for production of pure L-(+)-lactic acid. *Applied and Environmental Microbiology, 66*(9), 3835–3841. https://doi.org/10.1128/AEM.66.9.3835-3841.2000

Laopaiboon, P., Thani, A., Leelavatcharamas, V., & Laopaiboon, L. (2010). Bioresource technology acid hydrolysis of sugarcane bagasse for lactic acid production. *Bioresource Technology, 101*(3), 1036–1043. https://doi.org/10.1016/j.biortech.2009.08.091

Lapierre, L., Germond, J. E., Ott, A., Delley, M., & Mollet, B. (1999). D-Lactate dehydrogenase gene (ldhD) inactivation and resulting metabolic effects in the *Lactobacillus johnsonii* strains La1 and N312. *Applied and Environmental Microbiology, 65*(9), 4002–4007. https://doi.org/10.1128/aem.65.9.4002-4007.1999

Lee, J., Yup, S., Park, S., & Middelberg, A. P. J. (1999). Control of fed-batch fermentations. *Biotechnology Advances, 17*, 29–48.

Liu, B., Yang, M., Qi, B., Chen, X., Su, Z., & Wan, Y. (2010). Optimizing l-(+)-lactic acid production by thermophile Lactobacillus plantarum As.1.3 using alternative nitrogen sources with response surface method. *Biochemical Engineering Journal*, *52*(2–3), 212–219. https://doi.org/10.1016/j.bej.2010.08.013

Mamat, N. H., Mohd Noor, S. B., Che Soh, A., Saleena Taip, F., Rashid, A. H. A., Jufika Ahmad, N. L., Yusof, I. M., & Mohamad, A. Z. (2020). Soft-sensing estimation of optical density for pha production using multilayer perceptron neural network. *Journal of Physics: Conference Series, 1532*(1). https://doi.org/10.1088/1742-6596/1532/1/012019

Maqbool, M., Ali, A., Alderson, P. G., Mohamed, M. T. M., Siddiqui, Y., & Zahid, N. (2011). Postharvest application of gum arabic and essential oils for controlling anthracnose and quality of banana and papaya during cold storage. *Postharvest Biology and Technology, 62*(1), 71–76. https://doi.org/10.1016/J.POSTHARVBIO.2011.04.002

Maraveas, C. (2020). Production of sustainable and biodegradable polymers from agricultural waste. *Polymers, 12*(5). https://doi.org/10.3390/POLYM12051127

Mazumdar, S., Clomburg, J. M., & Gonzalez, R. (2010). *Escherichia coli* strains engineered for homofermentative production of D-lactic acid from glycerol. *Applied and Environmental Microbiology, 76*(13), 4327–4336. https://doi.org/10.1128/AEM.00664-10

Meng, Y., Xue, Y., Yu, B., Gao, C., & Ma, Y. (2012). Efficient production of l-lactic acid with high optical purity by alkaliphilic *Bacillus* sp. WL-S20. *Bioresource Technology, 116*, 334–339. https://doi.org/10.1016/j.biortech.2012.03.103

Montenegro, M. A., Boiero, M. L., Valle, L., & Borsarelli, C. D. (2012). Gum Arabic: More than an edible emulsifier. *Products and Applications of Biopolymers*. https://doi.org/10.5772/33783

Moon, S., Wee, Y., & Choi, G. (2012). A novel lactic acid bacterium for the production of high purity L-lactic acid, *Lactobacillus paracasei* subsp. paracasei CHB2121. *Journal of Bioscience and Bioengineering, 114*(2), 155–159. https://doi.org/10.1016/j.jbiosc.2012.03.016

Mosquera, L. H., Moraga, G., & Martínez-Navarrete, N. (2012). Critical water activity and critical water content of freeze-dried strawberry powder as affected by maltodextrin and arabic gum. *Food Research International, 47*(2), 201–206. https://doi.org/10.1016/J.FOODRES.2011.05.019

Naritomi, T., Kouda, T., Yano, H., & Yoshinaga, F. (2002). Influence of broth exchange ratio on bacterial cellulose production by repeated-batch culture. *Process Biochemistry, 38*.

Nolasco-hipolito, C., Zarrabal, O. C., Kamaldin, R. M., Teck-yee, L., Lihan, S., Bujang, K. B., & Nitta, Y. (2012). Lactic acid production by *Enteroccocus faecium* in liquefied sago starch. *AMB Express*, 1–9.

Odee, D. W., Telford, A., Wilson, J., Gaye, A., & Cavers, S. (2012). Plio-Pleistocene history and phylogeography of Acacia senegal in dry woodlands and savannahs of sub-Saharan tropical Africa: Evidence of early colonisation and recent range expansion. *Heredity, 109*(6), 372–382. https://doi.org/10.1038/HDY.2012.52

Oh, H., Wee, Y. J., Yun, J. S., & Ryu, H. W. (2003). Lactic acid production through cell-recycle repeated-batch bioreactor. *Applied Biochemistry and Biotechnology, 105*, 603–613.

Oun, A. A., & Rhim, J. W. (2017). Characterization of carboxymethyl cellulose-based nanocomposite films reinforced with oxidized nanocellulose isolated using ammonium persulfate method. *Carbohydrate Polymers, 174*, 484–492. https://doi.org/10.1016/j.carbpol.2017.06.121

Patel, M. A., Ou, M. S., Harbrucker, R., Aldrich, H. C., Buszko, M. L., Ingram, L. O., & Shanmugam, K. T. (2006). Isolation and characterization of acid-tolerant, thermophilic bacteria for effective fermentation of biomass-derived sugars to lactic acid. *Applied and Environmental Microbiology, 72*(5), 3228–3235. https://doi.org/10.1128/AEM.72.5.3228-3235.2006

Patel, S., & Goyal, A. (2015). Applications of natural polymer gum arabic: A review. *International Journal of Food Properties, 18*(5), 986–998. https://doi.org/10.1080/10942912.2013.809541

Penloglou, G., Chatzidoukas, C., & Kiparissides, C. (2012). Microbial production of polyhydroxybutyrate with tailor-made properties: An integrated modelling approach and experimental validation. *Biotechnology Advances, 30*(1), 329–337. https://doi.org/10.1016/j.biotechadv.2011.06.021

Pintado, J., Guyot, J. P., & Raimbault, M. (1999). Lactic acid production from mussel processing wastes with an amylolytic bacterial strain. *Enzyme and Microbial Technology*, 590–598.

Portnoy, V. A., Herrgård, M. J., & Palsson, B. (2008). Aerobic fermentation of D-glucose by an evolved cytochrome oxidase-deficient *Escherichia coli* strain. *Applied and Environmental Microbiology, 74*(24), 7561–7569. https://doi.org/10.1128/AEM.00880-08

Rachtanapun, P., Eitssayeam, S., & Pengpat, K. (2010). Study of carboxymethyl cellulose from papaya peels as binder in ceramics. *Advanced Materials Research, 93–94*, 17–21. https://doi.org/10.4028/www.scientific.net/AMR.93-94.17

Rachtanapun, P., Luangkamin, S., Tanprasert, K., & Suriyatem, R. (2012). Carboxymethyl cellulose film from durian rind. *LWT—Food Science and Technology, 48*(1), 52–58. https://doi.org/10.1016/j.lwt.2012.02.029

Rahman, M. S., Hasan, M. S., Nitai, A. S., Nam, S., Karmakar, A. K., Ahsan, M. S., Shiddiky, M. J. A., & Ahmed, M. B. (2021). Recent developments of carboxymethyl cellulose. *Polymers, 13*(8). https://doi.org/10.3390/polym13081345

Sabapathy, P. C., Devaraj, S., Meixner, K., Anburajan, P., Kathirvel, P., Ravikumar, Y., Zabed, H. M., & Qi, X. (2020). Recent developments in polyhydroxyalkanoates (PHAs) production—A review. *Bioresource Technology, 306*, 123132. https://doi.org/10.1016/j.biortech.2020.123132

Saber-Samandari, S., Saber-Samandari, S., Heydaripour, S., & Abdouss, M. (2016). Novel carboxymethyl cellulose based nanocomposite membrane: Synthesis, characterization and application in water treatment. *Journal of Environmental Management, 166*, 457–465. https://doi.org/10.1016/j.jenvman.2015.10.045

Salgaonkar, B. B., Mani, K., & Bragança, J. M. (2019). Sustainable bioconversion of cassava waste to poly(3-hydroxybutyrate-co-3-hydroxyvalerate) by *Halogeometricum borinquense* strain E3. *Journal of Polymers and the Environment, 27*(2), 299–308. https://doi.org/10.1007/s10924-018-1346-9

Santos, A. L., Francisco, M. G., Marques, S., & Roseiro, J. C. (2008). Lactic acid production from recycled paper sludge by simultaneous saccharification and fermentation. *Biochemical Engineering, 41*, 210–216. https://doi.org/10.1016/j.bej.2008.04.018

Saratale, G. R., Cho, S. K., Dattatraya Saratale, G., Kadam, A. A., Ghodake, G. S., Kumar, M., Naresh Bharagava, R., Kumar, G., Su Kim, D., Mulla, S. I., & Seung Shin, H. (2021). A comprehensive overview and recent advances on polyhydroxyalkanoates (PHA) production using various organic waste streams. *Bioresource Technology, 325*, 124685. https://doi.org/10.1016/j.biortech.2021.124685

Sasaki, M., Jojima, T., & Inui, M. (2008). Simultaneous utilization of D-cellobiose, D-glucose, and D-xylose by recombinant *Corynebacterium glutamicum* under oxygen-deprived conditions. *Applied Microbiology and Biotechnology*, 691–699. https://doi.org/10.1007/s00253-008-1703-z

Sasaki, M., Jojima, T., & Kawaguchi, H. (2009). Engineering of pentose transport in *Corynebacterium glutamicum* to improve simultaneous utilization of mixed sugars. *Applied Microbiology and Biotechnology*, 105–115. https://doi.org/10.1007/s00253-009-2065-x

Seddiqi, H., Oliaei, E., Honarkar, H., Jin, J., Geonzon, L. C., Bacabac, R. G., & Klein-Nulend, J. (2021). Cellulose and its derivatives: Towards biomedical applications. In *Cellulose* (vol. 28, Issue 4). Springer Netherlands. https://doi.org/10.1007/s10570-020-03674-w

Shibata, K., Flores, D. M., Kobayashi, G., & Sonomoto, K. (2007). Direct l-lactic acid fermentation with sago starch by a novel amylolytic lactic acid bacterium, *Enterococcus faecium. Enzyme and Microbial Technology, 41*, 149–155. https://doi.org/10.1016/j.enzmictec.2006.12.020

Shogren, R., Wood, D., Orts, W., & Glenn, G. (2019). Plant-based materials and transitioning to a circular economy. *Sustainable Production and Consumption, 19*, 194–215. https://doi.org/10.1016/j.spc.2019.04.007

Sirohi, R., Prakash Pandey, J., Kumar Gaur, V., Gnansounou, E., & Sindhu, R. (2020). Critical overview of biomass feedstocks as sustainable substrates for the production of polyhydroxybutyrate (PHB). *Bioresource Technology, 311*, 123536. https://doi.org/10.1016/j.biortech.2020.123536

Smith, A. M., Moxon, S., & Morris, G. A. (2016). Biopolymers as wound healing materials. In *Wound healing biomaterials* (Vol. 2). Elsevier Ltd. https://doi.org/10.1016/B978-1-78242-456-7.00013-1

Spierling, S., Knüpffer, E., Behnsen, H., Mudersbach, M., Krieg, H., Springer, S., Albrecht, S., Herrmann, C., & Endres, H.-J. (2018). Bio-based plastics—a review of environmental, social and economic impact assessments. *Journal of Cleaner Production, 185*, 476–491. https://doi.org/10.1016/j.jclepro.2018.03.014.This

Sreenath, H. K., Moldes, A. N. A. B., Koegel, R. G., & Straub, R. J. (2001). Lactic acid production by simultaneous saccharifkation and fermentation of alfalfa fiber. *Journal of Bioscience and Bioengineering, 92*(6).

Su, J., Flanagan, J., & Singh, H. (2008). Improving encapsulation efficiency and stability of water-in-oil-in-water emulsions using a modified gum arabic (Acacia (sen) SUPER GUM^{TM}). *Food Hydrocolloids, 22*(1), 112–120. https://doi.org/10.1016/J.FOODHYD.2007.03.005

Sudesh, K., Abe, H., & Doi, Y. (2000). Synthesis, structure and properties of polyhydroxyalkanoates: Biological polyesters. *Progress in Polymer Science (Oxford), 25*(10), 1503–1555. https://doi.org/10.1016/S0079-6700(00)00035-6

Surendran, A., Lakshmanan, M., Chee, J. Y., Sulaiman, A. M., Thuoc, D. V., & Sudesh, K. (2020). Can polyhydroxyalkanoates be produced efficiently from waste plant and animal oils? *Frontiers in Bioengineering and Biotechnology, 8*, 1–15. https://doi.org/10.3389/fbioe.2020.00169

Tashiro, Y., Kaneko, W., & Sun, Y. (2011). Continuous D-lactic acid production by a novel thermotolerant *Lactobacillus delbrueckii* subsp. lactis QU 41. *Applied Microbiology and Biotechnology*, 1741–1750. https://doi.org/10.1007/s00253-010-3011-7

Tu, W., Zhang, D., & Wang, H. (2019). Polyhydroxyalkanoates (PHA) production from fermented thermal-hydrolyzed sludge by mixed microbial cultures: The link between phosphorus and PHA yields. *Waste Management, 96*, 149–157. https://doi.org/10.1016/j.wasman.2019.07.021

Wang, Q., Zhao, X., Chamu, J., & Shanmugam, K. T. (2011). Isolation, characterization and evolution of a new thermophilic *Bacillus licheniformis* for lactic acid production in mineral salts medium. *Bioresource Technology, 102*(17), 8152–8158. https://doi.org/10.1016/j.biortech.2011.06.003

Wang, Y., Tian, T., Zhao, J., Wang, J., Yan, T., Xu, L., Liu, Z., Garza, E., Iverson, A., Manow, R., Finan, C., & Zhou, S. (2012). Homofermentative production of d-lactic acid from sucrose by a metabolically engineered *Escherichia coli. Biotechnology Letters, 34*(11), 2069–2075. https://doi.org/10.1007/s10529-012-1003-7

Wang, L., Xue, Z., Zhao, B., Yu, B., Xu, P., & Ma, Y. (2013). Jerusalem artichoke powder: A useful material in producing high-optical-purity L-lactate using an efficient sugar-utilizing thermophilic *Bacillus coagulans* strain. *Bioresource Technology, 130*, 174–180. https://doi.org/10.1016/j.biortech.2012.11.144

Wee, Y., & Ryu, H. (2009). Bioresource technology lactic acid production by Lactobacillus sp. RKY2 in a cell-recycle continuous fermentation using lignocellulosic hydrolyzates as inexpensive raw materials. *Bioresource Technology, 100*(18), 4262–4270. https://doi.org/10.1016/j.biortech.2009.03.074

Wee, Y.-J., Yun, J.-S., Kim, D., & Ryu, H.-W. (2006). Batch and repeated batch production of L (+) -lactic acid by *Enterococcus faecalis* RKY1 using wood hydrolyzate and corn steep liquor. *Journal of Industrial Microbiology*, 431–435. https://doi.org/10.1007/s10295-006-0084-5

Xu, K., & Xu, P. (2014). Efficient production of l-lactic acid using co-feeding strategy based on cane molasses/glucose carbon sources. *Bioresource Technology, 153*, 23–29. https://doi.org/10.1016/j.biortech.2013.11.057

Yukawa, H., Omumasaba, C. A., Nonaka, H., Watanabe, J., Ikeda, Y., & Verte, A. A. (2007). Comparative analysis of the *Corynebacterium glutamicum* group and complete genome sequence of strain R. *Microbiology*, 1042–1058. https://doi.org/10.1099/mic.0.2006/003657-0

Yustinah, H. N., Alamsyah, R., Roslan, A. M., Hermansyah, H., & Gozan, M. (2019). Production of polyhydroxybutyrate from oil palm empty fruit bunch (OPEFB) hydrolysates by Bacillus cereus suaeda B-001. *Biocatalysis and Agricultural Biotechnology, 18*, 101019. https://doi.org/10.1016/j.bcab.2019.01.057

Zhang, B., He, P., Ye, N., & Shao, L. (2008). Enhanced isomer purity of lactic acid from the non-sterile fermentation of kitchen wastes. *Bioresource Technology, 99*(4), 855–862. https://doi.org/10.1016/j.biortech.2007.01.010

Zhao, B., Wang, L., Ma, C., Yang, C., Xu, P., & Ma, Y. (2010). Repeated open fermentative production of optically pure L-lactic acid using a thermophilic Bacillus sp. strain. *Bioresource Technology, 101*(16), 6494–6498. https://doi.org/10.1016/j.biortech.2010.03.051

Zhou, S., Causey, T. B., Hasona, A., Shanmugam, K. T., & Ingram, L. O. (2003). Production of optically pure D-lactic acid in mineral salts medium by metabolically engineered *Escherichia coli* W3110. *Applied and Environmental Microbiology, 69*(1), 399–407. https://doi.org/10.1128/AEM.69.1.399-407.2003

Zhou, S., Yomano, L. P., Shanmugam, K. T., & Ingram, L. O. (2005). Fermentation of 10% (w/v) sugar to D(-)-lactate by engineered *Escherichia coli* B. *Biotechnology Letters, 27*(23–24), 1891–1896. https://doi.org/10.1007/s10529-005-3899-7

Zhu, Y., Eiteman, M. A., DeWitt, K., & Altman, E. (2007). Homolactate fermentation by metabolically engineered *Escherichia coli* strains. *Applied and Environmental Microbiology, 73*(2), 456–464. https://doi.org/10.1128/AEM.02022-06

Zou, H., Shi, M., Zhang, T., Li, L., Li, L., & Xian, M. (2017). Natural and engineered polyhydroxyalkanoate (PHA) synthase: Key enzyme in biopolyester production. *Applied Microbiology and Biotechnology, 101*(20), 7417–7426. https://doi.org/10.1007/s00253-017-8485-0

The manufacturer's authorised representative in the EU is Springer Nature Customer Service Centre GmbH, Europaplatz 3, 69115 Heidelberg, Germany. If you have any concerns regarding our products, please contact ProductSafety@springernature.com

Printed and bound by CPI Group (UK) Ltd, Croydon, CR0 4YY
01/10/2026
02225934-0005